THE CHIMPANZEES OF KIBALE FOREST

THE CHIMPANZEES OF KIBALE FOREST

A FIELD STUDY OF ECOLOGY AND SOCIAL STRUCTURE

MICHAEL PATRICK GHIGLIERI

Columbia University Press
New York   1984

Library of Congress Cataloging in Publication Data

Ghiglieri, Michael Patrick, 1946–
   The chimpanzees of Kibale Forest.

   Bibliography: p. 211
   Includes index.
   1. Chimpanzees.   2. Mammals—Uganda—Kibale Forest
Reserve.   3. Kibale Forest Reserve (Uganda)   I. Title.
QL737.P96G54   1984   599.88'44045'096761   83-10115
ISBN 0-231-05594-3

Columbia University Press
New York   Guildford, Surrey

Copyright © 1984 Columbia University Press
All rights reserved
Printed in the United States of America

Clothbound editions of Columbia University Press Books are Smyth-
sewn and printed on permanent and durable acid-free paper.

CONTENTS

LIST OF TABLES

I

LIST OF FIGURES

LIST OF PLATES

Plates appear as a group following page 106.

PLATE

1. Kibale Nature Reserve and Ruwenzori glaciers in background (photo by C. S. Ghiglieri).

2. Motorable track through primary forest southeast of Kibale Nature Reserve (photo by C. S. Ghiglieri).

3. Ngogo camp and a portion of the study area extending southwest beyond it (photo by C. S. Ghiglieri).

4. A typical view across a narrow belt of swamp forest at Ngogo (photo by M. P. Ghiglieri).

5. Swamp vegetation along the southern boundary of Kibale Nature Reserve (photo by M. P. Ghiglieri).

6. Buttresses of *Ficus mucuso* specimen #1 at Ngogo. I estimated its crown volume to be 21,303 m³, the largest at Ngogo (photo by M. P. Ghiglieri).

7. Elephants in a partially felled region immediately north of Kanyawara (photo by J. P. Skorupa).

8. Red duiker (photo by J. P. Skorupa).

9. Medan Mukasa, one of the Ugandan game guards assigned to Ngogo, with spears and snares confiscated after chasing poachers at Ngogo (photo by M. P. Ghiglieri).

10. Red colobus foraging on leaves (photo by L. A. Isbell).

11. Baboons in the forest at Ngogo (photo by C. S. Ghiglieri).

12. Gray-cheeked Mangabey at Ngogo (photo by M. P. Ghiglieri).

13. A redtail monkey leaping a gap in swamp forest at Ngogo (photo by M. P. Ghiglieri).

14. Solitary male blue monkey at Ngogo. Occasionally these usurp redtail harem masters and mate with female redtails to produce hybrids (photo by C. S. Ghiglieri).

15. Crowned Hawk-eagle in Kibale Forest Reserve, a predator on several species of primates (photo by C. S. Ghiglieri).

16. Talons of immature Crowned Hawk-eagle that apparently starved north of Kanyawara in Kibale Forest. The raptors apparently use these talons to pierce the skulls of their prey at first contact (photo by J. P. Skorupa).

17. A chimpanzee night nest (photo by M. P. Ghiglieri).

18. Chimpanzee in a day nest at Kanyawara (photo by J. P. Skorupa).

19. Adult male chimpanzee, Kong, at Kanyawara (photo by C. S. Ghiglieri).

20. Adult female chimpanzee, Gray, with a handful and mouthful of *Ficus mucuso* fig wadge at Ngogo. Very ripe figs were not wadged (photo by C. S. Ghiglieri).

21. Adult male chimpanzee, Eskimo, foraging in a *Ficus mucuso* at Ngogo (photo by C. S. Ghiglieri).

22. Subadult male chimpanzee, Fearless, and young adult female, Owl, foraging in a *Ficus mucuso* at Ngogo. Note size of Owl's perineum as sexual signal (photo by M. P. Ghiglieri).

23. Chimpanzees often rested during midday without building a day nest (photo by J. P. Skorupa).

ACKNOWLEDGEMENTS

The assistance of many institutions and individuals made this project possible, and I am grateful to all of them. The National Institutes of Health supported my transportation expenses from 1976–78 and salaries for game guards and porters from 1976–77 through NIH grant number MH 23008–04 to Thomas T. Struhsaker. The New York Zoological Society provided these expenses in 1978 and 1981. Professor G. Ainsworth Harrison of Oxford University and the Boise Fund awarded me a grant which supported six months of my project in 1978. The National Academy of Sciences awarded me a stipend grant that made possible three months of data analysis in 1978.

Thomas T. Struhsaker has my respect as a field biologist, conservationist and friend—and my thanks for his guidance, interest, and logistical support on this project. The friendship and hospitality of Ms. Lysa Leland enriched my life and assisted in my research in Kibale. Peter S. Rodman has my deep appreciation for his conception of this project and his unflagging enthusiasm, support, and guidance during all phases of it. I am grateful to William J. Hamilton III for originally suggesting that I study something smaller than a gorilla.

This research was possible only with the permission of the National Research Council, President's Office, Uganda, and the assistance of Mr. Frederick B. N. Mukiibi and Mr. X. K. Ovon. I am grateful also to the

Immigration Office of Uganda for expediting my entrance into their beautiful and hospitable country. Professor Ramsis Lutfy, Department of Zoology, Makerere University and Mr. Simeon Semakula, curator of the Makerere University Museum of Zoology, assisted in attaining my affiliation with the Department of Zoology. Mr. Anthony Katende, curator of the Makerere University Herbarium, assisted in identifying specimens of plant foods of chimpanzees in Kabale Forest. I am grateful to Mr. Peter Karani of the Uganda Forestry Department for his interest and official permission to conduct research in Kibale Forest. I am also indebted to Mr. Davis Semikole for his assistance during my first three days in Uganda. I am even more indebted to Oskar and Linda Rothen for their repeated warm hospitaliy in Kampala at short notice. I owe something to Simon Wallis for providing my initiation to the botany of Kibale Forest.

Lloyd and Fern Ghiglieri, my parents, handled my financial obligations in the U.S.A. and indulged me by procuring and mailing supplies I requested and luxuries I did not. Debra S. Judge took time and resources from her own field research in Baja California, and Linda Scott in Gilgil, Kenya did the same to send me logistical supplies and reminders of the West.

My special thanks go to Constance S. Ghiglieri for her assistance in proofreading and typing on drafts of this manuscript and for keeping me fed as I worked on it.

Last, but foremost, I wish to thank Conan M. Ghiglieri, my son, for understanding my two years of absence.

To other friends who have assisted me in this project, whose names I have fecklessly neglected to mention, I express my grateful appreciation.

Michael Patrick Ghiglieri
September 1983

THE CHIMPANZEES OF KIBALE FOREST

INTRODUCTION

In the eyes of four centuries of humanity, chimpanzees (*Pan troglodytes*) have embodied the intriguing combination of human-like morphology and beastial behavior. Along with their resemblance to humans, chimpanzees have demonstrated a remarkable intelligence and imitative ability that has softened their vague threat of evolutionary profanity with a clown-like air. Not until Goodall's (1971a) delightful *In the Shadow of Man*, though, did any appreciable portion of Western civilization have a glimpse of chimpanzee behavior in natural surroundings. Observations since then have prompted this book, the first report of a field study of wild, *unprovisioned* chimpanzees that discusses their ecology and social structure from an evolutionary perspective.

Not only are chimpanzees the closest living relatives of humans (as demonstrated by various lines of evidence, e.g., King and Wilson 1975; Sarich and Cronin 1976; Miller 1977; Fouts and Budd 1979), but also their social behavior approaches that of humans in its richness and complexity. Additionally, chimpanzees exhibit a rare and improbable constellation of ecological attributes; they are large-bodied, highly socialized, specialist frugivores adapted to terrestrial locomotion from arboreal ancestors. Their primary habitat is the tropical rain forest belt of Africa, a relatively stable natural laboratory and a habitat that, until this century, has afforded them sanctuary from severe competition from humans. For the theoretically minded ecologist and for those interested in the social structure and ecology of human and nonhuman primates, few other mammals evoke an interest equal to that of chimpanzees.

Despite the interest they engender and the several field projects that have attempted to clarify their social structure and ecology, our understanding of the nature of wild, undisturbed chimpanzees is unsatisfyingly incomplete. Early, short-term investigations (3–12 months) of chimpanzees (Nissen 1931; Azuma and Toyoshima 1962; Kortlandt 1962; Reynolds 1963; Sugiyama 1968, 1969; Albrecht and Dunnet 1971) yielded little or no data on the behavior, sociality, or ranging of individually identified chimpanzees, and their project designs tended to be ecologically unsophisticated. Connections between the type of social structure observed and the ecology of the apes often were ignored.

The two longitudinal studies of chimpanzees in Tanzania, at Gombe National Park (Goodall 1965, 1968, 1971a,b, 1973a,b, 1979 and many others) and at the Mahale Mountains (Nishida 1968, 1970, 1979, and others), have provided an excellent picture of individual behaviors. But because the observers gained the tolerance of the chimpanzees by providing them with permanently located, superabundant windfalls of unnatural domestic crops that were replenished daily for years, the subjects' normal social and ecological patterns were distorted (see Reynolds 1975). Wrangham (1974) noted that provisioning at Gombe promoted consistently large aggregations of individuals, and he warned, "Direct generalizations of the observational results gained at the feeding station to situations outside the feeding area should be avoided." Nishida (1979) admitted, "I do, of course, think that we should refrain from the provisioning method when doing purely ecological studies on a quantitative basis." Pusey (1979) suspected the provisioning at Gombe may have far-reaching effects upon the normal social development of maturing individuals. To further complicate interpretations, Sugiyama and Koman (1979) suggested that provisioning may have been responsible for female exogamy and male-retention. Riss and Busse (1979:291) concluded, "Unfortunately, the long history of banana feeding at Gombe has obscured attempts to assess the status of females as community members."

Salient features of chimpanzee sociality emerging from the Tanzanian studies were: a fusion-fission society, female exogamy (Nishida 1979;

Pusey 1979), which is a rare system in nonhuman primates but a common one among human societies (Murdock 1957), and apparent territorial maintenance by all male bands who occasionally engaged one another in border clashes (Bygott 1979). The latter phenomenon was particularly violent among the chimpanzees of Gombe (Goodall et al. 1979; Goodall 1979) and revealed an aspect of chimpanzee society that closely resembled primitive war among humans (see Vayda 1976).

Following a drastic reduction of an eight-year banana-provisioning program, the study community of Gombe chimpanzees split into two social units that became polarized in their ranging over a habitat they apparently once shared. Bands of males from the northern (Kasakela) community evinced a pattern of patrolling into the core area of the smaller, southern (Kahama) community. Over a period of two years, while traveling singly or with few companions, most of the Kahama adults were caught by the Kasakela males. Three adult males and two females were brutally killed; other individuals disappeared (Goodall et al. 1979). In addition, seven instances of infanticide were observed among the Kasakela chimpanzees, three by adult males directed against the infants of unfamiliar females who entered their range (Bygott 1972), and at least four fatal attacks on Kasakela infants by a Kasakela female (Goodall 1977).

Ecological trauma due to weaning the Gombe chimpanzees of their permanent banana windfall (44,000 calories per day, enough to meet the total daily metabolic needs of 14 adults; see Wrangham 1974) may have been exacerbated by an influx of chimpanzees from outside the park, where agriculture was displacing them. Fights between males of different communities occurred at Mahale Mountains and at Gombe, and in both places the combatants were individuals who had been fed regularly at the same provisioning station in each area. In light of the unknown but probably considerable influence of human activities at Gombe, the picture described above evades confident conclusions regarding the *normative* pattern (if a normative pattern exists) of chimpanzee sociality.

Ignoring for a moment the aspects of human-caused modifications to the ecological *modus vivendi* of the Gombe and Mahale chimpanzees, a theoretical model of social structure can be constructed from their behavior. Following the rash of primatological investigations that erupted in the equatorial regions of the earth in the 1960s came a trend to interpret naturalistic observations of primate behavior from an evolutionary perspective. This trend heralded an insightful phase of field ecology because it plumbed the roots of social structure by seeking to explain it as a complex result of adaptations to maximize the reproductive success of the social individual. Progress in this theoretical phase is characterized by much healthy argument (some of which is summarized in chapter 7), but it is generally agreed that social systems have evolved to maximize the reproductive success of the individuals in them—*within ecological and phylogenetic constraints*. From this perspective I will interpret the social system at Gombe.

The data indicating sexual differences among the Tanzanian chimpanzees lead to the following model of sexual selection (vide Darwin 1871) and territoriality among them. Using Wilson's (1971, 1975) criterion of exclusive use of space, the male chimpanzees are territorial. Considering the pattern of female exogamy, one may predict that a territorial boundary will hold a different significance for each age, sex, and reproductive class. If generations of males remain true to the territory of their natal community, then they will be more closely related to one another than to the immigrant females or to the average chimpanzee in their population (e.g., Glass 1953). Solidarity between these males is predictable on the basis of increased inclusive fitness resulting from kin selection (Hamilton 1964, 1972), although a model of reciprocal altruism (Trivers 1971) may also explain male–male cohesiveness in a territorial system as it explains the occasional partnerships between apparently unrelated male lions (*Panthera leo*), in the male exogamous system of the Serengeti lions described by Schaller (1972).

Whatever the balance of forces selecting for male–male cohesiveness, these adult males maximize their reproductive success by maintaining an exclusive home range, through the exclusion of alien males and their nonreproductive offspring, e.g., Bygott's (1972) and Suzuki's (1971) observations of infanticide. By maintaining the integrity of their territory to ensure exclusive access to its breeding females and to reserve its forage resources for themselves and their offspring, territorial males benefit through increased individual and inclusive fitness. But successful defense depends on traveling in parties of adequate size, so males should travel together both to repel alien males and to protect themselves from them.

Females will look at territorial boundaries differently. An estrous female maximizes her reproductive success by mating with an unrelated male (Sutter 1958; Reid 1976). She must leave her natal community to find such a mate. The permanence of her move is dependent upon her reception by the females of the new community (Goodall 1971b:92, 1977; Bygott 1979; Pusey 1979) with whom she must compete for food. Her decision to stay is probably also influenced by the quality of the new habitat. The males of the new community should want to mate with her and may try to force her to enter their territory (see Goodall et al. 1979). In short, to be evolutionarily successful, females must cross these boundaries.

A pregnant or postpartum female with an immature offspring, however, should avoid boundary areas because alien males in the vicinity may attack her and her offspring. When a desired resource in the peripheral region of the territory attracts such females with young, they should show keen vigilance to avoid being caught unaware.

This hypothetical model of social structure characterized by sexual selection and territoriality implies nonrandom membership in traveling parties and differences in the use of space by each age-sex-reproductive class. The hypothesis predicts:

(1) Adult males will be found traveling with other males more often than with any other age-sex-reproductive class (testable) because this enhances territorial defense. Goodall et al. (1979) and Nishida (1970, 1979) reported that parties of larger size generally were dominant during active intercommunity displacements.

(2) Adult males will apportion affiliative behaviors, such as mutual grooming, among other males in preference to other age-sex-reproductive classes (testable) because such behavior acts as social cement between individuals who may have to rely on one another during combat (not tested) and because they benefit through inclusive fitness (untested but logical).

(3) Adult males will spend more time traveling than will females with attendant offspring (testable) because the males must visit their territorial boundaries (untested), which encircle the entire community range, and must cover wide areas to locate females in estrous (untested but logical in the sense of Platt [1964]). A female needs to range only within an area of sufficient size to support herself and her attendant offspring.

The testable predictions of this hypothetical model may be tested by collecting data on traveling party membership, on mutual grooming, and on activity patterns. All of these data are collected more easily than observations of intercommunity conflicts, which, even when most feverish, are very rare events (Goodall et al. 1979).

Initially I planned to test this model at Gombe National Park, where intercommunity conflicts were taking on a striking importance in the lives of individual chimpanzees. Because I suspected that the patterns of ranging of individuals and the matrix of interactions between individuals among the original study community at Gombe were distorted by years of intensive provisioning of bananas, I planned to attempt a study of a new, unhabituated community to the north of Kasakela Valley. Convinced that behavioral data collected in an unnatural ecological context are of limited value, I intended, perhaps naïvely, to observe and habituate individuals of the new community *without* provisioning them. I was hoping to test with chimpanzees Schaller's (1963) technique of habituation: simply exposing oneself frequently but inoffensively. (He had employed this technique successfully with mountain gorillas, *Gorilla gorilla berengi*.) But before my plans passed beyond the proposal stage they were abrogated by political events.

In May 1975, a guerrilla task force from Zaïre, a unit of the reclusive Marxist Popular Revolutionary Party, slipped across Lake Tanganyika to Tanzania and staged a daring midnight raid on the Gombe research facility. They kidnapped three American students from Stanford University and a Dutch research assistant. Tanzania immediately closed Gombe to research by expatriates, and kept it closed even after the hostages were released. My proposed study became one of the scientific casualties.

Even though what had seemed to be an ideal situation for the research I planned had been lost, I remained intellectually committed to examining the problem of chimpanzee social structure. I needed a new study site, preferably one with an undisturbed habitat representative of chimpanzees as a species, rather than the more marginal savanna-woodland fringe along the east shore of Lake Tanganyika.

The three subspecies of chimpanzees (*P. troglodytes troglodytes, P. troglodytes verus,* and *P. troglodytes schweinfurthii*) and the pygmy chimpanzee (*Pan paniscus*) occur across a broad equatorial belt from Central Africa near 33°E. to the Atlantic coast and extending to about 13°N. and nearly 8°S. (see Reynolds and Reynolds 1965). Although such a wide distribution sounds promising on paper, in fact chimpanzees are restricted to remaining isolated pockets of suitable habitat, often merely small relict areas. In many places where they still occur they are hunted for food, which greatly reduces the number of potential study sites. Political and logistical difficulties narrow the choices even further. I was in a quandary over where to attempt a study.

Thomas T. Struhsaker convinced me that the Kibale Forest Reserve in western Uganda had potential as a study area for chimpanzees. In 1970 Struhsaker (1975, 1977, 1978), acting upon a suggestion by Peter Marler, had chosen Kibale as the site of his longitudinal studies of red colobus monkeys (*Colobus badius tephrosceles*) and redtail monkeys (*Cercopithecus ascanius*). Altogether eight species of anthropoids occurred there in sympatry. The habitat was a floristically complex mosaic of primary rain forest, montane forest associations, colonizing communities, and grassland. Local poachers concentrated their hunts to ungulates and elephants (*Loxodonta africana*); primates were not considered food. More than a year after the Gombe kidnappings, the President's Office of Uganda granted me permission to conduct field research there. Eventually I found that Struhsaker's assessment of Kibale as a feasible area for studying chimpanzees was correct.

My study was a systematic ecological investigation of chimpanzee sociology and ecology, unique because the apes were unprovisioned and because to some extent I compared the ecology and behavior of chimpanzees with that of sympatric anthropoids. The objectives of the project were:

(1) To determine the social structure of chimpanzees in Kibale Forest.

(2) To test the predictions generated by the hypothetical model of sexual selection and territoriality outlined above, and to note behaviors associated with intercommunity interactions.

(3) To determine the population density and structure of chimpanzees in the regions of Ngogo and Kanyawara in Kibale Forest Reserve.

(4) To determine food habits and the size, density, and distribution, and phenology of food patches and to relate these to party size and activity patterns of the chimpanzees. These data also facilitate an examination of niche separation with sympatric anthropods. The latter emphasis is aided by comparing data with previous and concurrent investigators of cercopithecid ecology in Kibale Forest.

This project was both theoretical and pragmatic. First, it was directed toward answering theoretical questions posed by the intriguing data collected during the longitudinal studies of provisioned chimpanzees in Tanzania. Second, it filled a gap in the available data on chimpanzee ecology and population density in a habitat of primary rain forest with no history of felling. Additionally, the data collected facilitate the analysis of chimpanzee sociality as a function of environment in the Type III sense of Altmann (1974)—i.e., by comparing the social structure and ecology of two closely related populations in different habitats.

Perhaps the major question raised by the Tanzanian episodes of territoriality is whether territoriality is a major social adaptation of chimpanzees in general. This was by no means clear, because other studies in Uganda and West Africa (Nissen 1931; Azuma and Toyoshima 1962; Kortlandt 1962; Reynolds 1963; Sugiyama 1968, 1969; Albrecht and Dunnet 1971), although short-term, did not suggest territoriality but just the opposite. Chimpanzee social structure was pictured as carnival-like with an open acceptance of strangers. In a habitat such as Kibale Forest, with an absence of unnatural, long-term and localized provisioning stations *that are defensible,* evidence of territoriality is likely to be more subtle (if it exists at all). However, verification of the predictions generated by the model of sexual selection and territoriality would be strong evidence in favor of a territorial interpretation of chimpanzee social structure and ecology in Kibale Forest.

The organization of this book follows what I considered to be natural divisions in the areas of investigation of chimpanzee social structure and ecology. Of course all chapters are interrelated. Although several might stand alone as articles in a scientific journal, together they provide more of a natural history of the apes. Omission of a chapter would leave a hole in the fabric of the picture and cripple the purpose of this book—to provide an evolutionary interpretation of the chimpanzees' *modus vivendi.* The final chapter contains a summary analysis of chimpanzee social structure and ecology in relation to that of the other great apes and of species more distantly related phylogenetically but exhibiting similar social systems.

Unfortunately, in a work of this type there is little space to convey the personal importance of the field work and its ramifications. My experience in Uganda was both diverse and obsessively goal oriented—and uneven thanks to the vagaries of Idi Amin Dada's government. But my life in the forest enriched me in ways that can never be lost. I became adapted to its natural rhythm and tuned to the lives of its denizens. I became an insider in a microcosm that had long been shunned by civilized man as a dangerous jungle. Above all, though, my eventual acceptance by the chimpanzees and the data they allowed me to collect will always remain the highlight of my career.

1

KIBALE FOREST RESERVE

Although chimpanzees occur in many habitats typified by the presence of mature, fruit-bearing trees, most naturalistic observations of these apes have been made in areas chosen for their convenience to the observer: the peripheries of agricultural plantations, rain forests disturbed by partial or patchy felling, and the broad woodland fringes that separate the true rain forest from the savanna. Viewed within the perspective of the current geographical range of chimpanzees and what is known of their feeding ecology, none of these areas can be considered habitats typical of chimpanzees as a species. In contrast, the central and southern sections of Kibale Forest appear vegetatively similar to the vast rain forest ecosystem to the west, the core of the natural geographic range of chimpanzees.

The Kibale Forest Reserve has been described in detail by Kingston (1967, not generally available), Wing and Buss (1970), and Struhsaker (1975). The following summary description of the forest as a whole primarily reflects the latter two references.

The Kibale Forest Reserve is located in the Toro District of western Uganda (figure 1) southwest of the Uganda Plateau of the Great Rift Valley and approximately 24 km east of the Ruwenzori Mountains. Almost reaching the equator, its geographic coordinates are 0° 13′ to 0° 41′ N and 30° 19′ to 30° 32′ E (figure 2). Its 56,000 ha. extend along an

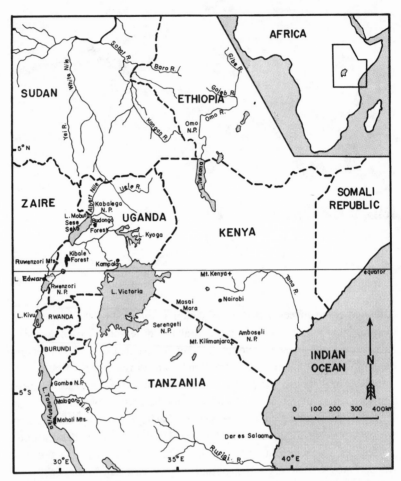

FIGURE 1. East Africa.

approximately 36 km north to south tilt from 1,590 m to 1,110 m elevation broken by well-watered, undulating series of hills and valleys whose variations in relief rarely exceed 150–180 m vertically (Wing and Buss 1970). The warm and rainy climate allows for streams and swamps in most valleys, with all drainage ultimately occurring via two southerly directed watercourses, the Dura and Mpanga rivers, which drain into Lake George (Wing and Buss 1970; Struhsaker 1975).

The vegetation of the Kibale Forest Reserve forms a complex edaphic and topographic mosaic of grassland, woodland thicket, and colonizing forest (40%) and tropical forest types (60%) in various seral stages and states of disturbance (Wing and Buss 1970). In general, grassland, woodland thickets, and similar vegetative associations forming early seral stages of forest succession occur in areas of higher relief that once were covered by forest, reflecting recent volcanic activity (Osmaston 1959) and subsequent habitation and cultivation by humans. A virulent epidemic of *Ngana* (sleeping sickness) sweeping across western Uganda early in the twentieth century forced the local Batoro tribesmen out of the major portion of Kibale Forest. At present the early seral stages are apparently continuing in a forest succession, but due to continued burning by poachers, change only very slowly and hence represent a fire disclimax. Within the remaining 60 percent of the reserve, plant associations resembling moist evergreen forest, related to moist montane forest and lowland tropical deciduous forest—all characterized by conspicuous dominants—form a floral cline along the north–south axis of the reserve (Struhsaker 1975).

Economic value of Kibale Forest derives principally from the value of trees as timber. Felling and marketing of trees has progressed under a 70-year felling cycle planned in 1948 when the region was designated a Central Forest Reserve and is proceeding from north to south (Kingston 1967). The geographic occurrence of high-value timber within the reserve is extremely patchy, with species appearing, becoming dominant, then disappearing as one walks for a few hours toward the equator. The pattern of timber extraction has reflected this patchiness and has created blocks of disturbed habitat adjoining primary forest.

Local avian and mammalian fauna are diverse, reflecting Kibale Forest as an interface habitat containing wildlife typical of Central and East Africa. Elephant and nine species of ungulate are present, African buffalo (*Syncerus caffer*), bushbuck (*Tragelaphus scriptus*), sitatunga (*Tragelaphus spekei*), red duiker (*Cephalophus harveyi*), blue duiker (*Cephalophus monticola*), waterbuck (*Kobus defassa*), bushpig (*Potomachoerus porcus*), giant forest hog (*Hylochoerus meinertzhageni*), and warthog (*Phacochoerus aethiopicus*). Lions and hippopotami (*Hippopotamus amphibius*) occurred in the reserve historically but apparently have been extirpated. Recent poaching activity has reduced elephant and buffalo drastically (see Eltringham and Malpas 1976; Van Orsdol 1979) and also may be responsible for only token representation of waterbuck and warthog.

Leopard (*Panthera pardus*) signs are few. Along with spotted hyena (*Crocuta crocuta*), golden cat (*Felis aurata*) and a variety of viverids (civits, genets, and mongooses) are present.

In addition to chimpanzees, seven other diurnal species of primates are sympatric within the reserve, baboons, (*Papio anubis*), redtail monkeys, blue monkeys (*Cercopithecus mitis*), hybrid individuals of *C. ascanius* and *C. mitis,* l'Hoest's monkeys (*Cercopithecus l'hoesti*), gray-cheeked mangabeys (*Cercocebus albigena*), red colobus, and black and white colobus (*Colobus guereza*).

Prior to 1962, both Batoro and Bakonjo tribesmen hunted game (illegally) in Kibale Forest. For the Bakonjo game meant many birds and most large mammals, primates included. Baboons and chimpanzees are still considered a delicacy by the Bakonjo, and it is probably safe to assume that within the period of this study there lived both chimpanzees and baboons old enough to remember having been hunted (see section on chimpanzee interactions with humans). During a civil war in 1962, the Batoro defeated and drove the Bakonjo into the Ruwenzori Mountains. Batoro consider primates as unfit for human consumption so they are no longer hunted in Kibale Forest for that purpose. Batoro poachers continue to set and maintain lines of wire snares within the forest for trapping ungulates. Reports by local people indicate that chimpanzees sometimes were trapped by these and were occasionally killed by poachers.

KANYAWARA

I visited Kanyawara (compartment 30 of the Kibale Forest Reserve, figure 2) for brief periods as a secondary study area to Ngogo (described below). Although only 6.6 percent of my time in the field was spent at Kanyawara, the area was important because it facilitated observations (about 10 percent of my total) of a community of chimpanzees separate from that which frequented Ngogo.

The Kanyawara study area is a northern-jutting peninsula of primary rain forest surrounded on the west and north by grassland, exotic softwood plantations (planted with *Pinus caribaea, P. radiata, Cupressus lusitanica* and *Eucalyptus* sp.) and small relict stands of primary rain forest, and to the east primarily by grassland.

South to southeast extends the rain forest. The study area is a trail-grid

KIBALE FOREST RESERVE, UGANDA

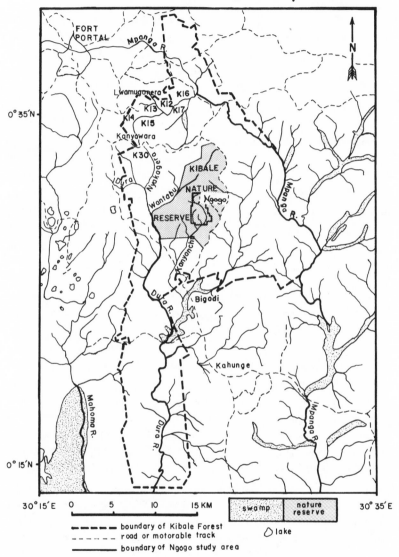

FIGURE 2. Kibale Forest Reserve, Uganda.

system of 1.5 km² in essentially the headwaters of the Nyakagera creek, a narrow valley bottom of swamp forest bordered by steep forested slopes leading upward to end abruptly in a narrow forest-grassland colonizing ecotone. The grasslands are dominated by stands of elephant grass (*Pennisetum purpureum*), which attain heights of up to 3 m and offer stiff resistance to one traveling on foot. Compartment 30 was the site of Struhsaker's longitudinal investigation of red colobus and is described in greater detail in Struhsaker (1975).

In addition to compartment 30, I observed chimpanzees in compartment 14, adjacent to and northwest of it. This area was similar to compartment 30 but more level and it had been selectively felled. Predominant among trees left standing were *Ficus sp.*, especially *F. natalensis* and *F. exasperata*, both of which produced fruit utilized by the Kanyawara community of chimpanzees. I conducted a chimpanzee nest count in this compartment and also in compartments 12, 13, 16, and 17 (see figure 2) along a forest access road.

NGOGO, THE MAJOR STUDY AREA

When Kibale Forest was assigned as a forest reserve, Ngogo was set aside as a small (approximately 2 km²) nature reserve, an inviolate status second only to the security of a national park. Although a large portion of the nature reserve was grassland dominated by elephant grass, felling operations for timber extraction never penetrated within roughly an 8 km radius of the nature reserve itself—so, in essence, Ngogo was the formally recognized core of a much larger tract of forest. In 1975 the Kibale Nature Reserve was greatly expanded by a far-sighted and conservation-minded ruling by the Uganda Forestry Department, and as a result it is more likely to remain a viable habitat (figure 2). Because of its relatively undisturbed state (people have not lived in the region since the 1930s), Struhsaker (1977) chose Ngogo as a secondary study area for investigating red colobus and redtail monkeys. For this study of chimpanzees Ngogo was the primary study area.

Ngogo is located 10 km southeast of Kanyawara and is connected to it by a 12 km game trail from the northwest and a motorable track from the south. The study area (0° 29' to 0° 31' N; 30° 24' to 30° 26' E) extends southwesterly through a tract of primary rainforest beginning on the western edge of an irregularly shaped series of elephant-grass dominated hills in the central region of Kibale Forest between 1280–1420 m elevation. The study area proper is approximately 6.5 km², 0.5 km² of which is dominated by elephant grass distributed in two hilltop enclaves surrounded by primary forest (figure 3). The remaining 6 km² of forest has

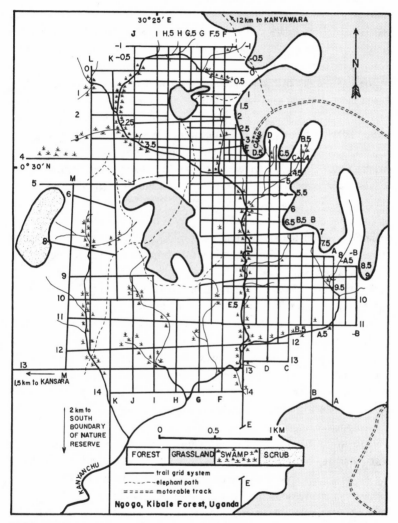

FIGURE 3. Ngogo study area, Kibale Nature Reserve, Uganda.

been divided into cells approximately 50 m, 100 m, or 200 m on a side (the smallest cells being closest to camp) by a north–south and east–west trail-grid system approximately 85 km in total length. The marked trails facilitated orientation, ability to map position, and rapid mobility. An additional 6 km of trail were cut as seven extensions from the study area south toward the southern boundary of the nature reserve and west toward Kansara. In addition to these cut trails, I used the game path between Ngogo and Kanyawara and the boundary lines cut by the Forestry Department to delineate the northern and southern extent of Kibale Nature Reserve for census routes.

The topography of Ngogo is hilly, with well-forested slopes extending to the narrow forest-grassland ecotone. Floral associations on the slopes tend to form an intricate series of clines apparently related to drainage and/or edaphic factors not yet understood. Wherever valley bottoms are not strictly v-shaped, a typical open-canopied floral association known as swamp forest occurs. In wide valley bottoms swamps were dominated by sedges such as papyrus (*Cyperus papyrus*) and other herbaceous forms interrupted by an occasional tree such as *Ficus dawei, Neobutonia macrocalyx,* or *Symphonia globulifera,* typical of swamp forest. Swamp associations within the study area are a minor plant community occurring primarily along wide valley bottoms of the main stem of the south-flowing Kanyanchu stream, whose headwaters extend from the northern edge of the study area. Swamp forest is a more extensive and widespread community but covers only a small fraction of the forest proper.

Daily records were kept of rainfall and maximum and minimum air temperatures. A rain gauge on the cut sward in camp, approximately 50 m from the edge of the forest, was checked and emptied around 0700 hours, and the maximum-minimum thermometer, approximately 50 m inside the forest, was checked and reset next. These data were collected by a Ugandan assistant, and I checked them periodically for accuracy.

Precipitation at Ngogo was 1,664 mm during 1977, a 12.8 percent increase over the 52-year annual average precipitation reported for the Fort Portal region by Wing and Buss (1970) (figure 4). Precipitation during the first 522 days of the study occurred on 223 days or 42.7%. Four distinct annual seasons are characteristic of this region; a rainy season from March through May and another from September through November, with two drier seasons occurring between these. Most days were

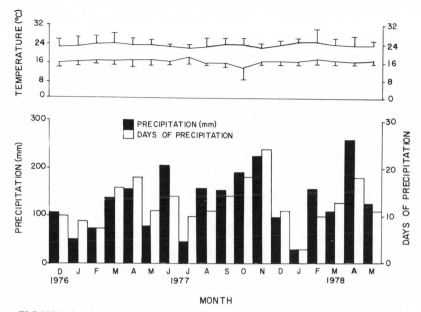

FIGURE 4. Meterological data from Ngogo, December 1976–May 1978. Vertical lines designate extreme temperature range per month. Lines indicate mean high and low temperatures per month. Total precipitation, 2,336 mm.

overcast, even during the dry seasons when cloud cover was exacerbated by smoke from recurrent agricultural and grassland fires. Temperatures were moderate with differences between extreme maximum and minimum temperatures rarely exceeding 12°C during most months (figure 4).

All of the mammals described above as being present in Kibale Forest were present in Ngogo during my study of chimpanzees with the possible exception of leopard. (Game guards have reported kills cached in trees but I found no evidence of presence of leopard at Ngogo.) Anomalies were noted in regard to habitat utilization by two species of primates. North of trail #2 no groups of black and white colobus were seen, although they did occur further north of Ngogo. South of trail #9 no blue monkeys were seen (figure 3). Ngogo apparently straddles the southernmost range for blue monkeys in Kibale Forest. I saw chimpanzees in all portions of the study area except grassland and papyrus swamp.

2

INVESTIGATIONAL TECHNIQUES

I directly observed chimpanzees in Kibale Forest for 488 hours between 9 December 1976 and 17 September 1977, between 29 October 1977 and 14 May 1978, and between 6 January and 31 May 1981. In December 1977 I also observed chimpanzees in Budongo Forest (table 1).

THE CHIMPANZEES

Habituation

The popular notion that chimpanzees will not become habituated to the presence of an observer unless offered an incentive, such as a large concentration of fruit or other food that they find difficult to resist, made the habituation aspect of my study a challenge. Because the primary focus of this study was ecological, I was determined to introduce nothing new (other than myself as an observer) into the system I wanted to study. Schaller (1963) was successful in habituating groups of mountain gorillas simply by showing himself to them repeatedly and inoffensively. This approach is less effective with chimpanzees because they range further and travel in smaller, inconsistent parties. Each encounter by an observer produces exposures with fewer chimpanzees and fewer exposures with the same chimpanzee. My initial contacts with Ngogo chimpanzees left me with the unequivocal conclusion that they had no tolerance for humans.

TABLE 1. HOURS OF OBSERVATION AND ENCOUNTERS WITH CHIMPANZEES PER MONTH AT NGOGO AND KANYAWARA BETWEEN 9 DECEMBER 1976 AND 14 MAY 1978 AND BETWEEN 6 JANUARY AND 1 JUNE 1981.

Month	Ngogo				Kanyawara			
	Observation of chimpanzees (hr:min)	Number of contacts with chimpanzees	Active research in forest (hr:min)	Chimpanzee encounters per hour in forest	Observation of chimpanzees (hr:min)	Number of contacts with chimpanzees	Active research in forest (hr:min)	Chimpanzee encounters per hour in forest
December 1976	0:01	1	80:00[a]	0.01	0	0	10:00	0
January 1977	37:24	185	180:00[a]	1.03				
February 1977	15:25	115	150:00[a]	0.77				
March 1977	62:15	202	180:00[a]	1.12				
April 1977	14:32	74	150:00[a]	0.49				
May 1977	27:28	107	150:00[a]	0.71				
June 1977	10:15	23	152:46	0.15	3:04	19	9:35	1.98
July 1977	27:32	64	214:31	0.30				
August 1977	15:03	21	153:02	0.14				

September 1977	19:54	19	81:46	0.23				
October 1977	0:15	3	15:37	0.19				
November 1977	27:42	83	111:29	0.74				
December 1977[b]	1:49	9	60:10	0.15	8:26	16	40:55	0.39
January 1978	22:02	47	216:04	0.22				
February 1978	7:59	20	87:22	0.23				
March 1978	18:53	89	188:57	0.47				
April 1978	67:28	229	182:33	1.25				
May 1978	19:05	53	89:25	0.59				
January 1981					0	0	21:30	0
February 1981					0	0	19:15	0
March 1981	0	0	73:57	0	15:06	34	32:58	1.03
April 1981	3:09	17	137:11	0.12	24:28	96	56:29	1.70
May 1981	38:31	124	173:37	0.71	0	0	7:30	0
					0	0	4:38	0
TOTAL=	436:42	1,485	2,828:27	0.53	51:04	165	202:50	0.82

[a] Estimate only

[b] During December an additional 10 hours and 20 min were spent in Budongo Forest resulting in 40 min of observation on 17 chimpanzees.

By stationing myself in obscure, but not hidden positions approximately 50 m from large fruiting *Ficus* spp. I was able to make repeated contacts with several chimpanzees. I observed individuals as they arrived at a tree, determined their age-sex-reproductive classes and, eventually, their identities, usually before they noticed me. During the initial phase of habituation, chimpanzees always fled when they detected me. Often, however, the same individuals returned to a tree several more times during its fruiting stage. They learned where to look for me upon entering a tree and fled again once they saw me. After seeing me a few times some individuals rushed *through* a tree in what I thought of as "lightning fig raids," picking and stuffing figs into their mouths and hands to carry away for later consumption. As such raids produced no unpleasant consequences, some individuals were prompted to extend their time in a tree, staying to eat for up to an hour. Eventually the same individuals remained to nest, to groom themselves or another chimpanzee, to play, etc. In short, several individuals became habituated to me. During this habituation phase I engaged in innocuous activities, such as exaggerated scratching or eating *Afromomum* leaves, and I was careful not to threaten the apes by staring at them unwaveringly.

Some solitary chimpanzees took longer than others to get over their fear of me. Of two adult males who often associated with one another, Raw Patch tolerated me in less than a month, while Silverback became tolerant only after 15 months. Responses toward me by individuals in social groupings varied depending on its members. Habituated individuals acted as catalysts in allaying the fears of less tolerant chimpanzees. Some individuals remained in my presence only if certain others present were apparently unconcerned. In general, adult females with infants were the least tolerant age-sex-reproductive class, while adult females with large perineal swellings were the most tolerant. Once individuals were habituated, I could walk about freely beneath them in a tree without disturbing their behavior. Some individuals in the Ngogo community never habituated to me; I sometimes approached these by stealth and concealed myself during observation.

Most of the chimpanzees of the Kanyawara community were tolerant of humans in a nonspecific way. Perhaps because of many years of contact with workers near Kanyawara Forestry Station, even infants were habituated to my presence almost immediately. The Kanyawara community ranged through a habitat which in part has been severely disturbed by forest management and timber felling operations; for this reason the chimpanzees of Ngogo were my primary study community.

Identification and location

I maintained an identification file of individual chimpanzees at Ngogo and Kanyawara with written descriptions of characteristics useful for their identification and with some drawings of their faces. I named individuals when I considered my ability to identify them reliable, or, in the case of infants and dependent juveniles, when I could identify their mothers. With most adults the time required for learning their identities was one hour of good observation.

All life stages of chimpanzees herein are my estimates. The characteristics I used are listed in table 2. The actual ages in years listed in table 2 derive entirely from the longitudinal work of Goodall (1968, 1971a) and Nishida (1968), who have observed the same individuals for the greatest number of years.

My observations of chimpanzees were dependent upon my locating them. In order of decreasing success my three primary methods for finding chimpanzees were: (1) all-day or part-day "vigils" at large trees with ripe fruit (a vigil is basically the same as the habituation technique described above), (2) obtaining a compass bearing on the location of vocalizing chimpanzees and then searching for them in that direction (the success of this method was limited because chimpanzees often vocalized while traveling and were gone when I reached the area indicated by the compass bearing), and (3) randomly walking the trail-grid system and looking for chimpanzees. On some days I used a combination of all three approaches. Whenever I watched a chimpanzee build a night nest during the evening I tried to arrive at the nest early the next morning to resume observations. Sometimes, however, the apes left their nests before ambient light was adequate for observation.

Data recorded

Sampling behaviors. Once contact had been made with one or more chimpanzees I noted the date, then recorded the number and age-sex-reproductive class of the individuals present, their identities, and their order of progression (if a traveling party). During subsequent systematic

TABLE 2. PHYSICAL CHARACTERISTICS USED TO SUBJECTIVELY ESTIMATE AGES OF CHIMPANZEES IN KIBALE FOREST, UGANDA.

Life stage	Age (years)	Color of hair	Color of face, hands and feet	Males			Females		
				Body weight (kg)	Sexual characters	Social characters	Body weight (kg)	Sexual characters	Social characters
Infant	0–4	Black, tail tuft white.	Pink.	≤10	Conspicuous penis, no testicles visible.	Carried by mother, no younger siblings.	≤10	Undeveloped, clitoris visible.	Carried by mother, no younger siblings.
Juvenile	4–9	Black. Long white tail tuft gradually reduced to nothing.	Pinkish-tan to blotchy dark tan.	10–30	None or inconspicuous testicles.	Often have younger siblings and may travel with adult males.	10–25	None or slight perineal swellings.	Often have younger siblings and often travel with mother.

Stage	Age	Males	Weight (lb)	Males	Males	Age	Females	Females
Subadult	9–14 (males only)	Black, some brown or lighter tints. No tail tuft.	30–40	Testicles well-developed. Musculature well-developed.	Usually have younger siblings. May travel with any age-sex class.	25–35	Not really comparable stage to males. Females become rounder, darker in skin, lighter in hair. May have "permanent" perineal swellings. May mate but nulliparous.	
Adult	≥15 males ≥11 females	Brown, sometimes blond or black. Muzzle gray to white. Back may be lighter in color against lumbar and sacral region.	40–50	Testicles very large. Musculature sometimes massive.	Similar to above. Often dominant over all other age-sex classes.	30–43	Very large perineal swellings during estrus. Maternal teats.	Often are accompanied by immature, dependent offspring.

NOTE: Body weights are my estimates. Ages for life stages derive from Goodall (1968, 1971a) and Nishida (1968).

sampling the following data were recorded instantaneously at 5-min intervals: the time, location and height above ground (if appropriate) of focal animal, tree species occupied and food type if feeding, activity and posture, nearest conspecific neighbors of whom I was aware, distance to sympatric primates if in interspecific association, and any response or activity directed toward a nonchimpanzee neighbor, including myself. Interspecific and intraspecific interactions, defecation or urination, and/or vocalizations were recorded continuously. Weather was noted when transitions occurred from one type to another. The activity or activities recorded at each 5-min interval were those in which the animal was engaged at the ending of a 10-sec countdown ending at the 5-min point.

Approximately halfway through the study I began recording the first activity engaged in by the focal animal which lasted ≥5 sec during each 5-min interval. This additional datum was recorded to facilitate interobserver, interspecific comparisons within Kibale Forest. Initially T. Struhsaker suggested recording activity data of ≥5-sec duration rather than instanantaneous sampling. I found that the 5-sec constraint eliminated activities of shorter duration, such as scratching, yawning, some vocalizing, picking fruit, etc., which I considered to be equally important as scanning, chewing, etc. By lumping data from both instantaneous and 5-sec duration samples into the gross activity categories, foraging, resting and traveling, however, I found they were virtually identical. The advantage of instantaneous sampling is that a more diverse and complete inventory of activities is recorded. The activities in this report are instantaneous data.

During my initial contact with a party I usually chose the first chimpanzee I saw as focal animal. I tried to retain the same focal individual for as long as possible. When that individual left the party or became hidden for 15 or 20 min I switched to a new focal animal, usually the one most clearly visible to me. When I had a choice between sexes I tried to focus on the one least well represented in my recent data. In practice some individuals were focal individuals for only one 5-min interval while others were observed continuously for up to 9 hours.

Occasionally I opportunistically recorded social interactions between individuals not including the focal animal. This was feasible in situations when the focal individual was engaged in a single, continuous activity, such as sleeping in a nest, that was easily monitored.

Instances of pant-hooting, a long-range call (Goodall 1968; Marler and Hobbet 1975), were recorded whenever I heard them (between 21 April 1977–10 May 1978). Data recorded included date, time, estimated location and number of vocalizers, and whether or not buttress-drummings accompanied the vocalizations. When distant pant-hooting coincided with my observation of chimpanzees, I recorded the vocal response or nonresponse of those individuals under observation.

Spatial monitoring. The Ngogo study area is divided into cells by a trail-grid system. Trail junctions are labeled with signs made from squares of corrugated iron sheeting lettered by pounding nail holes through them, then painting over the dotted figures. Each trail is marked at 50 or 100 m intervals using the same system so that one can determine a precise location within the study area. Chimpanzees rarely traveled along the trail-grid system; they apparently preferred the more direct route afforded by game trails. When I followed traveling parties I usually was able to use trail coordinates to record their locations accurately.

The majority of my observations were restricted to localized areas of the forest. This was especially true during vigils at fruit trees. Most chimpanzees at Ngogo did not tolerate me to the extent that I could walk along behind them as they traveled, so my daily ranging data were difficult to collect. Follows on individuals who tolerated me sometimes were hampered by the forest itself. The understory vegetation was normally so thick as to impede my bipedal locomotion, and it limited horizontal visibility to a radius of about 10 m. Unless a focal individual cooperated by waiting for me to catch up, I was unable to stay with it. In fact, none cooperated. My follows were limited to less than a full day over distances of 1.5 km or less. On some days I recontacted focal individuals without knowing their intervening itineraries.

Population density

Direct measure by line-transect censuses. I conducted censuses of chimpanzees and other species of large mammals along six fixed routes. At Ngogo I used a 5.5 km route in the southern portion ($N = 24$), a 3.4 km route in the north ($N = 22$), a 3.0 km route in the southeast ($N = 3$), a 5.5 km route along the southern boundary of Kibale Nature Reserve ($N = 3$), and a 5.0 km route along a game path from Ngogo to and including part of the northern boundary of Kibale Nature Reserve ($N =$

3). In addition I twice censused a 6.5 km route from compartment 30 to the northern boundary of the nature reserve (see figure 5).

Censuses usually began between 0730 and 0800 hours, often on consecutive days of each month. Data recorded for each sighting of a large mammal included the route, date, weather, time, location, height above

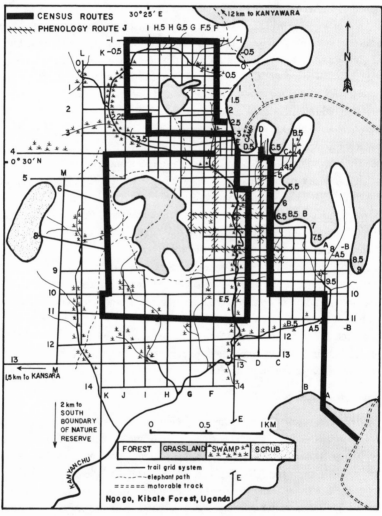

FIGURE 5. Census and phenology routes at Ngogo.

ground of animal (if appropriate), horizontal sighting distance between me and the first individual sighted of each social grouping, food species and type, approximate or actual size of social grouping, and the nature (if any) of interspecific association. Interspecific associations are defined as any occasion when an individual of one species is surrounded by members of another species. When two "solitary" individuals of different species were within 25 m of one another I considered them to be in association. Usually this meant they were in the same tree. Recognizable vocalizations of animals not seen were also recorded, with an estimate of the number of vocalizing individuals and their location, but these data proved too vague for analysis. My average rate of progress during censuses was 1 km/hour.

The key data used to determine the area sampled are the horizontal sighting distances from my position on the census route to the first member seen in each social grouping. Data from a large sample may be analyzed to determine the maximum reliable sighting distance. This distance is one half the width of the strip sampled during the census and varies between species. The number of individuals or groups counted within the maximum reliable sighting distance during all censuses is divided by the area sampled (2 × maximum reliable sighting distance × total length of trail sampled) to yield an estimate of density.

Indirect measure by enumeration of nests. Chimpanzees constructed nests by pulling one or more boughs inward to form a crude platform within the crotch of a tree or upon a nearly horizontal limb. Sometimes the crotch of one tree was used for support while the platform and padding materials were pulled in from as many as four other trees. Often these secondary branches were broken off completely and added to the nest apparently to improve its comfort. This final process sometimes continued at sporadic intervals during the use of the nest.

Chimpanzees built some nests for night use and others as resting spots during the day. The latter nests were built quickly, half took less than 1 min to construct, and frequently were so scanty that they became unidentifiable as nests immediately after use, in contrast to night nests which appeared substantial. But a few day nests appeared similar to and were indistinguishable from night nests on the basis of their construction.

Early reports on nest building (e.g. Goodall 1965) stated that each chimpanzee built a new nest every night. However, several investigators, including Goodall (1965), Kortlandt (1962), Izawa and Itani (1966), and Wrangham (1975), mention reuse of old night nests. Goodall (1968) suggested that the reuse of 20 nests at Gombe during a 4-year period may have been associated with an abnormal density of chimpanzees around the banana-provisioning area such that all the best sites had nests in them already and so were reused. I observed a chimpanzee at Ngogo reuse an old night nest only once. Reynolds and Reynolds (1965) speculated that some chimpanzees past infancy did not build a nest at all on some nights. Kortlandt (1962) made the same speculation. Neither spent nights in the forest monitoring chimpanzees, though. Adults at Ngogo sometimes fed well into darkness and then resumed feeding at dawn, thus giving the impression that they had been doing so all night. Whether or not they constructed night nests is unknown.

Izawa and Itani (1966) reported super-large nests they considered to have been built to rest two or more chimpanzees, although they never saw more than one in a nest. All of these deviations suggest that the ratio of one chimpanzee past infancy in the forest per one new nest built each night may not be accurate. In view of their rarity, though, the deviation may be trivial. On the other hand, a few elaborate day nests may easily be mistaken for night nests during an enumeration of the latter. Despite these complications my impression is that on the average over time each chimpanzee past infancy built approximately one nest which appeared to be a night version per 24-hour period.

Although it was not possible to test quantitatively, I suspect that reuse of old night nests was less prevalent than my misidentification of elaborate day nests as night nests. So, while the ratio might be unity, my ability to be certain of each nest's previous use probably led to a total count of night nests slightly higher than what actually existed, and my computations of chimpanzee density based on nest counts probably are biased upward.

But knowing the ratio was approximately 1 to 1 makes it possible to estimate the population density of chimpanzees from the density of nests in an area. Requisite data include the lifespan and density of nests in the area and the population structure of chimpanzees.

I recorded the location and date of construction of nests I saw being built or nests whose appearance I noted the following morning at a site I had visited the previous evening. I determined their lifespan by monitoring 28 nests to gauge the time they remained recognizable as such. Recognizability was subjective; if I no longer was confident in stating that a nest existed where it had before, it was not recognizable.

I obtained an estimation of night nest density by using a line-transect technique to count all night-type nests within 10 m of either side of the center of the trail within the trail-grid system. I made recounts along a few trails to determine my efficiency in locating nests. This was about 80 percent. The population structure of the nest builders became evident as I came to know the chimpanzees individually. The formula below incorporates the data discussed to estimate the average density of chimpanzees over time in the area of the nest count:

$$\frac{1}{\text{chimpanzee nest/day}} \times \frac{\text{Number of nests counted}}{\text{Area sampled (km}^2)} \times \frac{1}{\text{Observer efficiency}} \times$$

$$\frac{1}{\text{Average life-span of nest}} \times \frac{\text{Number of chimpanzees}}{\text{Number of nest builders}} = \text{Chimpanzees/km}^2$$

A similar technique was improvised by Schaller (1961) to estimate the density of orangutans (*Pongo pygmaeus*) in Sarawak, although the dimensions of observer efficiency (100%), life span of nests (6 months), and number of nests built by each ape over time (1/day) all were assumed values. Rijksen (1978) discussed the threefold range of values assumed by several workers to be the period during which orangutan nests remain recognizable and their untested assumption that each ape beyond infancy built only one nest each day. Rijksen found that each Sumatran orangutan built 1.8 nests/day and the average nest remained visible for 76 days.

In addition to numbers of nests counted within the 20 m-wide strips in Kibale Forest, I recorded the height, location, tree species, and relative age of each nest. However, because it was not possible to ascertain whether nests in a clump were contemporaneous, I did not use these data to determine social structure (see Casimir 1979).

ECOLOGICAL CORRELATES

Opportunistic observations of sympatric mammals

In addition to the systematic data recorded during censuses of mammals, I made efforts to record ecological data on rarely seen species or rarely observed events among the more common species. The former category includes l'Hoest's monkeys, blue monkeys, hybrid blue–redtail monkeys and baboons, all of which I saw in association with chimpanzees and feeding on the same food types. Encounters with elephants, buffaloes, other ungulates, mammalian predators, and snakes were recorded because little is known of these animals' habits in Kibale, though connections with chimpanzee ecology are not clear. Opportunistic observations on feeding competition, group sizes of sympatric mammals, and interspecific associations were recorded to aid in constructing a picture of competition with chimpanzees.

Availability of major food species

Phenology. The patterns of availability of primary food species may place constraints upon the ranging of chimpanzees and, to a certain extent, upon their socialization (e.g., Izawa 1970; MacKinnon 1974; Wrangham 1975). To sample the distribution of food in time I monitored the phenology of 107 specimens of 14 species of tree known or suspected to be major food species of the Ngogo chimpanzees. Of the 107 specimens, 28 trees were part of a continuing phenology series initiated by S. Wallis. The remaining 79 trees, mainly *Ficus* spp., I chose because of my early experience with the chimpanzees. Of these trees, 67 were located along a phenology route (figure 5), 34 were along my southern census route, and 6 were along the northern census route in the trail-grid system.

I monitored these trees once each month, the latter two groups during censuses of mammals, at intervals of as near to one month as possible. T. Struhsaker regularly sampled 30 phenology trees on a separate route, 15 of which were food species of chimpanzees. During 5 months of his absence during 1977–78, I monitored these in addition to mine. Characteristics monitored for each tree included quantity and ripeness of fruit, quantities of young and/or mature leaves and flowers, and extent of damage to foliage. Each characteriatic was scored on a relative scale of 0 to 4, with 4 being the maximum quantity possible for a particular tree. Estimates of the actual numbers of fruits on any tree were not attempted because I was unable to find a practical technique that accurately assesses the quantity of fruit within crown volumes of up to 21,000 m^3 beginning more than 10 m above the forest floor.

Patch size. Most of the focal trees of fruit tree vigils attracted successive parties of chimpanzees. I estimated the crown volumes of 25 vigil trees to examine the relationship between patch size and numbers of chimpanzees who exploited the patches. I measured the horizontal distances from the northern to the southern extremes of the crown by taking compass bearings perpendicular to the north–south and east–west oriented grid trails to those extremes, then by marking the site along the trail where the orientation to the crown extreme was 90°. The two marks on each trail, representing opposite extreme extensions of the crown, limit a line parallel to the actual distance across the crown and equal it. I estimated the height of the crowns by eye. Averaging the two halved horizontal measures of crown width with the estimate of height yielded an estimate of crown radius, assuming a hemisperical shape.

Because the growth forms of the tree crowns were roughly hemispherical, computation of the volume of each crown was simplified by using the formula $\frac{2}{3}\pi r^3$. The crown volumes were not exactly equal to the figures generated by the formula because of lacunae peculiar to their growth forms. But because I found no striking differences between crowns to indicate that lacunae were more extensive in one tree than another, the estimation of radii and crown volumes provides a valid *relative* index of patch size useful for making the examination of patch size in relation to size of feeding parties of chimpanzees.

Patch dispersion. The density and distribution in space of food species influence social patterns of chimpanzees. During the first two months of this study I mapped *Ficus* spp. within 10 m of each side of the center of all trails within the trail-grid system at Ngogo to gain an estimate of density and distribution. All specimens of some very large species visible from *any* part of a trail, e.g., *Ficus mucuso* and *F. dawei,* were mapped to gain an idea of their distributions and to scout for possible vigil trees.

T. Struhsaker conducted a line-transect enumeration of all trees ≥ 10 m tall within 2.5 m of each side of the center of the trail along 17.35 km by the end of this project at Ngogo. He received slight assistance from S. Wallis and me. The region of the Ngogo study area sampled was the home range of Struhsaker's (1977) study group of redtail monkeys. It is not known whether the area was typical of Ngogo as a whole, but these unpublished data are useful for relative abundance of forage species used by chimpanzees.

Topographic features and mapping

During the same scouting walk along the trail-grid system during which I enumerated *Ficus* specimens and chimpanzee night nests, I also mapped the trail-grid system. I noted the locations of topographic features such as watercourses, swamps, hilltop grasslands, and large sections of colonizing scrub.

Unless otherwise noted (as above) all original data presented in this report were collected by me without assistance.

3

POPULATION STRUCTURE, DENSITY, RANGING AND DYNAMICS

Obtaining accurate data on the composition and size of groups of rain forest primates is a notoriously difficult task (Struhsaker 1975; Cant 1978a; Rodman 1978:471). Struhsaker (1975:181) estimated, "it usually requires about 60 hours of contact to obtain a complete count of a small group of red colobus and considerably longer for large groups." The advantage of working on a rain forest primate exhibiting a discrete and compact group structure is that such complete counts are possible at all. Determining the membership of a community of chimpanzees is even more difficult. They frequently traveled or fed in temporary parties which rarely remained stable for more than a day. Group or community structure among these apes is only revealed over time as the product of all regular interindividual associations within their home range. At no time, or in a single location, did I see all of the individuals who together formed the social matrix defined here as the "community" (c.f. Reynolds and Reynolds 1965; Goodall 1968; Sugiyama 1968; Nishida 1970, 1979; and see chapter 7). Hence no classic "group count" was possible.

COMMUNITY SIZE AND POPULATION STRUCTURE

Community size

The first prerequisite to delineating community size and structure is determining the identities of the resident individuals. The use of known individuals circumvents some of the guesswork faced by Cant (1978b) in determining the age-sex structure of groups of spider monkeys (*Ateles geoffroyi*), a species exhibiting an unusual fusion-fission society much like that of chimpanzees.

At Ngogo I recognized 46 individuals (listed with field descriptions in Appendix A). All but three of these, a single matrilineal family, were seen interacting as part of the Ngogo community. They may have also been part of the community. Because I saw them only once, and then without others nearby, their status is unclear. I saw some of the identified adult females without offspring during only one month of the study, and, because of this, I suspect they also may not have been regular community members. In addition to the 46 identified individuals, I saw at least 8 to 12 other chimpanzees of various age-sex classes whom I did not see well or often enough to reliably identify. All of these factors considered, I estimated the Ngogo community to contain about 55 members.

I spent too little time in Kanyawara to identify all community members, so no reliable estimate of the size of that community is possible here. I identified 24 associating individuals there (see Appendix B for descriptions). Traveling parties there were more cohesive than their counterparts at Ngogo. From this, and from the similar appearance of the membership of the parties I observed in Kanyawara, I would as a guess estimate that the community size there is about 35 individuals. The cohesive party structure during travel was similar to that seen among small communities of chimpanzees in Filabanga by Kano (1971:238), in Kabogo and Ishanda by Itani and Suzuki (1967), and in Mahale Mountains by Nishida (1968:178).

The size of chimpanzee communities apparently varies between and within habitats. In the savanna-woodland habitat of Tanzania Itani and Suzuki (1967) reported 21, 31, and 43 individuals; Kano (1971) reported 12–15 and 23; Nishida (1968, 1979) reported 29 and 80; and Riss and Busse (1977) reported 42 for the Kasakela community of Gombe. In contrast, Reynolds (1963) estimated community size at approximately 70 individuals and Sugiyama estimated it at about 50 in Budongo Forest. Though admittedly based on limited data, a trend is evident for larger communities inhabiting rain forests ($\bar{x}$ = 58.3 chimpanzees) than more thinly wooded areas ($\bar{x}$ = 35.7 chimpanzees). This may be due to greater population densities of the apes in rain forests (see section on population density below).

Population structure

The criteria I used to estimate the ages and determine the sexes of Kibale chimpanzees are described in table 2. Standardization of life stages (age-sex-reproductive designations and age limits) among chimpanzees is lacking between workers (c.f. Reynolds and Reynolds 1965; Goodall 1968; Nishida 1968; Sugiyama 1968; Izawa 1970; Albrecht and Dunnett 1971). Their designations follow the classic polarization between ''splitters'' and ''lumpers'' depending on the relative importance each worker attached to age classes such as ''baby'' ($\leq$0.5 years) as opposed to ''infant'' (0.5–2.0 years) (Sugiyama 1969:198). I adopted a lumper classification based on the major life history stages of the animal. Infants range only in the company of their mothers. Juveniles often range with their mothers and siblings but may range solitarily, with siblings only, or with parties containing no mothers and possibly no siblings. Subadult males range with other adult males but are reproductively inferior to them. Adult males range primarily with each other. Adult females range primarily in units consisting of mothers and their dependent offspring, or in parties with other adult females, although adults of both sexes traveled with individuals of all life stages at some time. These are statistically significant age-sex-reproductive classes (see chapter 6) correlated with different life history stages. The group structures of the Ngogo and Kanyawara communities are summarized in table 3.

Although the ratio of males to females across all age classes was unity among Kibale chimpanzees, the ratio among adults and subadults was 0.88. Male:female ratios among adults varied from other studies: 0.35 in the Kasakati Basin (Itani 1967), 0.6 in Mahale Mountains (Nishida 1968), 0.6 at Gombe (Riss and Busse 1977), 1.29 at Budongo (Sugiyama 1968), and 0.67 at Bossou, Guinea (Sugiyama and Koman 1979). Excluding the study at Budongo, all observations indicate a greater number of adult or nearly adult females in chimpanzee communities. Sex ratios at birth ap-

TABLE 3. AGE-SEX CLASS REPRESENTATION IN TWO COMMUNITIES OF CHIMPANZEES IN KIBALE FOREST, UGANDA.

Study area	Sex	Age class					
		Adult ≥15 yrs	Subadult 10–15 yrs	Juvenile 5–9 yrs	Infant 0–4 yrs	Total	Percent
Ngogo (May 1978)	Male	10	3	4	4	21	45.7
	Female	13	4	4	2	23	50.0
	Unknown	–	–	–	2	2	4.3
	Subtotal	23	7	8	8	46	100.0
Kanyawara (May 1981)	Male	7	1	2	2	12	50.0
	Female	6	1	2	1	10	41.7
	Unknown	–	–	1	1	2	8.3
	Subtotal	13	2	5	4	24	100.0
	Total	36	9	13	12	70	
	Percent	51.4	12.9	18.6	17.1	100.0	

pear near unity. Three factors may account for this inequity in adulthood: (1) females reach maturity a year or two sooner than males (Goodall 1968); (2) females emigrate to new communities from their natal community and may reside in a new community only temporarily (Nishida and Kawanaka 1972; Pusey 1979) so that, during a long study, temporary females may be added with permanent females and compared with the number of permanent males, who do not emigrate, thus indicating an unrealistic sex ratio for the population as a whole; and (3) males engage more frequently in displays and intercommunity combats (at least at Gombe and Mahale) and may be subject to differential mortality by accidental death (Teleki 1973) or as a result of combat (Goodall et al. 1979).

Fully mature individuals accounted for 51.4 percent of the population sampled. This proportion of the Kibale population is near the mean for those reported in other areas: 65.0 percent (with no subadult females listed among 22 individuals older than juveniles) at Kasakati basin (Itani and Suzuki 1967), 51.7 for the Kajabara group at Mahale (Nishida 1968), 44.6 at Budongo Forest (Sugiyama 1968), and 57.1 at Gombe (Riss and Busse 1977).

POPULATION DENSITY

Line-transect censuses for densities of primates

Maximum reliable sighting distance. Line-transect censuses were the primary means of estimating the population densities of chimpanzees and other primates in the region of Ngogo. One thorny problem associated with generating estimates of density from such censuses is the determination of the real width of the strip sampled (see Janson and Terborgh, no date). Emlen (1971), Robinette et al. (1974) and Cant (1978a) used the perpendicular distance from the path of the observer to the animal sighted as a figure equal to one half the sample strip width. Struhsaker (1975) argued that the distance between the observer and the first animal sighted in a social group was more representative of the true area sampled. Cant (1978a) equated the perpendicular, horizontal distance from the path of the observer with the sighting distance. Less rigorous, Albrecht (1976) used line-transect censuses in Budongo Forest but reported chimpanzee density simply in the dimensions, contacts per hour. For this study I considered the sighting distance to be the most relevant measure of sample strip width, but the maximum reliable sighting distance for each species must be computed from the entire pool of detection distances.

Visibility in the rain forest was limited. Because of the differential behavior, coloration, and grouping propensities of each primate species, they differed in their detectability. Table 4 presents the variability in species-specific detection distances and indicates the maximum reliable sighting distance for each. For social groups of most arboreal primates this was 50 m; for solitary monkeys and groups of terrestrial monkeys it was much less. Width of the strips sampled was a species-specific value equal to double the maximum reliable sighting distance. The area sampled was this figure multiplied by the total length of the line transect.

Numerical density. Table 5 presents results of 55 censuses along five routes of varying length totaling 242 km. Because too few repetitions were made along each route for them to be analyzed individually and because the coefficients of variation for numbers of sightings per species along each were similar, I pooled all data pertaining to density from all routes to yield an estimate for Ngogo and Kibale Nature Reserve as a whole. The northernmost census route, which was sampled twice, was outside the reserve and was not included in the computations of density below. These censuses also may represent chimpanzee density in primary forest habitat for the entire central block of Kibale Forest, but many repetitions along routes distant from Ngogo would be required to ascertain this. Only sightings which fell within the maximum reliable sighting distance were used in the computations of densities in table 5.

Initially I suspected that the northern census route at Ngogo was scarce in fauna relative to the others. The northern habitat was steeper with a more dense understory and more broken canopy. *Ficus dawei* and *F. mucuso* were rare in the north but common in the south, while large specimens of *Parinari excelsa,* rare in the south, were common in the north. In the north I rarely saw black and white colobus, who are habitat specialists in Kibale (Oates 1974), but censuses revealed that other primates were equally represented there. No two of the five census routes analyzed seemed to sample exactly the same combination of floral associations and, in general, I saw more chimpanzees along the three southern

Species	Social grouping	Distance (m) from observer when first sighted								
		0–10	11–20	21–30	31–40	41–50	51–60	61–70	71–80	>81
Chimpanzee	Party		1	4	5	5.			1	
Baboon	Group		3	3	5.	2	1	1		
	Solitary		2.	1	1					
Redtail monkey	Group	2	11	22	31	21.	4	2	1	1
	Solitary	1	2	7	6.	1				
Blue monkey	Group			3	.	2.				
	Solitary			1						
l'Hoest's monkey	Group				1.	1				
	Solitary		1		1.					
Gray-cheeked mangabey	Group			7	10	7.	1			1
	Solitary			1	1.	1	1			
Red colobus monkey	Group		1	7	16	19.	9	8	1	6
	Solitary				1.	1				
Black and white colobus monkey	Group			4	3	6.		1	2	
	Solitary			1		2.				
Bushpig	Party	1	1.							
Giant forest hog	Solitary	1		1						
Red duiker	Solitary	19	41.	15	4					
Blue duiker	Solitary	6	9.	1						
Bushbuck	Solitary	1	8.		2					

NOTE: The maximum reliable sighting distance for each species and social grouping is designated by a period.

TABLE 5. RESULTS OF 55 CENSUSES ALONG 5 LINE-TRANSECT ROUTES OF VARYING LENGTHS.

Species	% of all censuses containing species	Range in sightings per census	X̄ no. of sightings per km of census	X̄ no. of social groups per 100 ha (1 km²)	X̄ no. of solitaries or individuals per 100 ha (1 km²)
Primates					
Chimpanzee	25.5	0–3	0.066	0.026 community	1.45 individuals
Baboon	29.1	0–2	0.078	0.57	0.21 solitaries
Redtail Monkey	85.5	0–5	0.458	3.51	0.77
ascanius-mitis hybrid	1.8	0–1	0.004	n/a	
Blue Monkey	7.3	0–1	0.016	0.12	0.05
l'Hoest's monkey	7.3	0–1	0.016	0.05	0.10
Gray-cheeked mangabey	50.9	0–4	0.165	1.28	0.15
Red colobus	80.0	0–4	0.343	1.90	0.05
Black & white colobus	34.5	0–2	0.086	0.54	0.15
Ungulates					
Bushpig	3.6	0–1	0.008		0.52 individuals
Giant forest hog	3.6	0–1	0.008		0.31
Red duiker	81.8	0–7	0.388		7.95
Blue duiker	27.2	0–3	0.078		1.86
Bushbuck	23.6	0–3	0.062		1.03

NOTE: Lengths were 5.5 km (N = 2), 3.4 km, 3.0 km, and 2.6 km, totaling 242 km. Censuses were conducted between December 1976 and May 1978 and January–May 1981 in Kibale Nature Reserve.

routes ($\bar{x} = 0.07$ contacts/km) than during the three northern routes ($\bar{x} = 0.05$ contacts/km).

At an estimated density of 1.45 individuals/km^2, chimpanzees are one of the rarest primates in the forest. On an individual basis redtail monkeys outnumbered them 85 to 1. All 7 species of monkeys collectively outnumbered the apes by 197 to 1. Other rare primates included l'Hoest's monkeys and blue monkeys (south of compartment 30). The only ungulates the censuses indicated as plentiful were red duikers (7.95/km^2), a finding that agreed with my impression during noncensus periods.

Because of the dense understory in most areas, it was sometimes impossible to identify ungulate species even when closer than 20 m. Among the primates, l'Hoest's monkeys, baboons, and chimpanzees sometimes were not visible on the ground beyond 20 m unless in conspicuous social groups. Because questionable sightings were not included in the computation of density, the estimates in table 5 are conservative.

Animals such as elephant, cape buffalo, and waterbuck were not included because they were rarely present and because the census routes sampled only small portions of the study area. During 3031 hours of research in Kibale Forest, I saw elephant only on four occasions, buffalo twice, and waterbuck once. These large herbivores were under poaching pressure and acted wary in the presence of humans. Warthogs were present in the grasslands at Ngogo but not in the forest. Nocturnal animals including spotted hyena, civets, genets, mongooses, and prosimians (*Galago* spp.) also were present in the forest but not in censuses.

A comparison of the estimates of primate density in table 5 with Struhsaker's (1975:290–292) estimates of density in compartment 30 reveals differences in numerical representation by species in habitats separated by only 10 km. Baboons, common residents at Ngogo (approximately 1 troop/2 km^2), are absent from compartment 30. Chimpanzees and mangabeys are more numerous at Ngogo, while blue monkeys and black and white colobus are far less numerous. My estimate of redtail density is also lower at Ngogo, but the most dramatic difference between the two study areas is in the numbers of red colobus present. Struhsaker estimated 5.95 groups/km^2 in compartment 30, while my estimate for Ngogo was only 1.9 groups/km^2. Struhsaker also counted only one third as many groups of red colobus per linear km of transect in the area of the Dura River Bridge (essentially the same as my census route along the southern boundary of Kibale Nature Reserve) than in compartment 30, which suggests that, from the perspective of red colobus at least, there exist significant differences in habitat types between the two regions. Large-crowned fig trees, for instance, apparently are more common at Ngogo, where species of frugivorous primates predominate.

Nest counts for density of chimpanzees

Life span of nests. In order to interpret the density of nests of chimpanzees in terms of their builders the average life span of nests must be known. I monitored 28 nests of known date of construction, in the six most frequently used species of trees, on a monthly basis until each was no longer recognizable as a nest. I considered the age of each to be the period from the date of construction to the date midway between the last monthly inspection when it was recognizable and the first when it was not. Factors determining life spans were: (1) gradual recuperation of the tree, (2) dismantling of nests by gray-cheeked mangabeys (S. Wallis, pers. comm.) and by blue monkeys searching for insects within, (3) building a new nest on top of an old one, and (4) destruction by tree falls.

Table 6 presents the results of nest monitoring. Nests lasted from 15 to 202 days and, considering all tree species pooled, lasted an average of 110.8 days. Because of the wide variance in mean life spans between species of nest trees and because chimpanzees used certain species predominately and unequally (see below), I used a weighted mean for each nest-count–based estimate of chimpanzee density in table 9 (below) considering those species whose average life spans were known from table 6. For species not monitored for life span I used the estimate of 110.8 days. Many previous workers using nest counts to estimate population densities of apes (Schaller 1961; Kano 1972; Yoshiba 1964) assumed that nests lasted an average of 6 months.

Nest enumeration. I enumerated nests along census routes, along other trails cut between felled compartments 15, 14, 12, 13, and 17 northwest of Kanyawara, and along the southern boundary of Kibale Nature Reserve south of Ngogo.

Table 7 summarizes my three nest counts at Ngogo; of 451 nests 93.9 percent were constructed in 5 of the 21 identified species of trees used. Those top 5 species constituted 52.8 percent of all the 86 species enu-

TABLE 6. SUMMARY OF DETERIORATION RATES OF CHIMPANZEE NESTS.

Tree species	No. of specimens monitored	Range in life span (days)	S.D. (days)	$\bar{X}$ no. of days recognizable as a nest
Uvariopsis congensis Rabyns & Chesquiere	7	15–158	49.9	95.3
Chrysophyllum albidum G. Don	6	90–201	38.8	143.5
Diospyros abyssinica (Hiern.) F. White	6	15–195	64.4	119.2
Monodora myristica (Gaertn.) Dunal	6	45–202	68.7	120.8
Celtis durandii Engl.	3	62–91	16.7	71.7
Ficus mucuso Welw. ex Ficalho	1	N/A	N/A	80.0
Total	28[a]	15–202	54.3	110.8

NOTE: Rates are determined by monitoring nests of known age constructed in the six most frequently used tree species at Ngogo, Kibale Forest, Uganda. The date upon which a nest was considered to be no longer recognizable was the date midway between the date last recognized and subsequent date when classified as unrecognizable.
[a] Total is 28 rather than 29 because specimen No. 19 was a nest constructed using limbs from Chrysophyllum a. and Diospyros a. and was included in the separate analyses for both species.

TABLE 7. SUMMARY OF UTILIZATION OF TREE SPECIES FOR NIGHT NESTS BY NGOGO CHIMPANZEES.

Tree species	Whole nest count[a]				% representation in total nest sample				Range in heights of nests (m)	Average height (m)
	(a)	(b)	(c)	(d)	(a)	(b)	(c)	(d)		
Uvariopsis congensis	145.5	33.0	21.0	10.0	39.1	67.3	70.0	47.6	2–20	7.7
Chrysophyllum albidum	97.5	2.0	2.0	1.0	26.2	4.1	6.7	4.8	2–25	14.3
Diospyros abyssinica	33.5	6.0	5.0	0	9.0	12.2	16.7	0	7–25	14.3
Monodora myristica	32.5	0	2.0	7.0	8.7	0	6.7	33.3	8–23	15.4
Celtis durandii	24.5	1.0	0	0	6.6	2.0	0	0	6–27	15.3
Ficus mucuso	5.0	1.0	0	0	1.3	2.0	0	0	23–35	29.8
Pterygota mildbraedii	6.0	0	0	0	1.6	0	0	0	13–32	21.6
Piptadeniastrum africanum	5.0	0	0	0	1.3	0	0	0	15–33	28.2
Funtumia latifolia	1.0	2.0	0	0	0.3	4.1	0	0	13–20	17.0
Parinari excelsa	1.0	2.0	0	0	0.3	4.1	0	0	25–35	29.3
Strombosia scheffleri	3.0	0	0	0	0.8	0	0	0	9–14	11.3
Olea welwitshii	2.0	0	0	0	0.5	0	0	0	16–32	24.0
Pseudospondias microcarpa	2.0	0	0	0	0.5	0	0	0	10–16	13.0
Mimusops bagshawei	1.5	0	0	0	0.4	0	0	0	22–30	26.0
Cordia millenii	1.0	0	0	0	0.3	0	0	0	n/a	32.0
Ficus brachylepis	0	1.0	0	0	0	2.0	0	0	n/a	11.0
Teclea nobilis	0	1.0	0	0	0	2.0	0	0	n/a	6.0
Warburghia Ugandensis	1.0	0	0	0	0.3	0	0	0	n/a	7.0
Premna angolensis	0	0	0	1.0	0	0	0	4.8	n/a	8.0
Aphania senegalensis	0.5	0	0	0	0.1	0	0	0	n/a	15.0
Symphonia globulifera	0.5	0	0	0	0.1	0	0	0	n/a	16.0
Unidentified species	9.0	0	0	2.0	2.4	0	0	9.5	2–30	11.9
Total	372.0	49.0	30.0	21.0	99.8	99.8	100.1	100.0	2–35	12.2

NOTE: Nest enumerations were made along: (a) 58.2 km of trail during January–August 1977, (b) 9.68 km of trail during January–February 1978, (c) 10 km of trail during March 1981, and (d) 5.5 km of trail in March 1981. All enumerations were made along census and other trails in Kibale Nature Reserve, Uganda.

TABLE 8. SUMMARY OF UTILIZATION OF TREE SPECIES FOR NIGHT NESTS BY KANYAWARA CHIMPANZEES.

Tree species	Whole nest count[a]			% representation in total nest sample			Range in heights of nests (m)	Average height (m)
	(a)	(b)	(c)	(a)	(b)	(c)		
Uvariopsis congensis	37.0	30.5	14.0	58.7	63.5	37.8	5–15	8.4
Diospyros abyssinica	11.0	8.0	10.0	17.5	16.7	27.0	5–21	11.4
Celtis durandii	5.0	0	0	7.9	0	0	7–19	15.2
Tedea nobilis	3.0	1.0	0	4.8	2.1	0	6–10	8.0
Aphania senegalensis	0	3.0	0	0	6.2	0	6–11	9.0
Mimusops bagshawei	2.0	1.0	0	3.2	2.1	0	17–23	20.7
Strychnos mitis	0	2.0	0	0	4.2	0	5–6	5.5
Bosqueia phoberos	1.5	0	0	2.4	0	0	10–12	11.0
Pancovia turbinata	0	1.5	0	0	3.1	0	5–8	6.5
Chrysophyllum gorungosanum	1.0	0	0	1.6	0	0	n/a	10.0
Dombeya mukole	1.0	0	0	1.6	0	0	n/a	20.0
Spathodea campanulata	0	1.0	0	0	2.1	0	n/a	6.0
Strombosia scheffleri	0.5	0	3.0	0.8	0	8.1	6–10	9.0
Premna angolensis	0	0	2.0	0	0	5.4	6–7	6.5
Markhamia platycaylx	0	0	1.0	0	0	2.7	n/a	15.0
Olea welwitshii	0	0	1.0	0	0	2.7	n/a	10.0
Unidentified	1.0	0	6.0	1.6	0	16.2	5–18	9.1
Total	63.0	48.0	37.0	100.1	100.0	99.9	5–23	10.8

NOTE: Nest enumerations were made along: (a) 10.09 km of trail during December 1977, (b) 9.37 km of trail during May 1978, and (c) 9.03 km of trail during March 1981 in compartment 30 at Kanyawara, Kibale Forest Reserve, Uganda.
[a] Proportion of nest made up by species.

merated by Struhsaker (unpublished data) in a sample of 3227 trees at Ngogo. At Kanyawara I enumerated nests along a route including T. Struhsaker's 4.0 km census route (see table 8). Of 148 nests 82.8 percent were constructed in 5 of the 14 identified species of trees used. These top 5 species constituted 45.1 percent of all the 51 species enumerated by Struhsaker (1975) in a sample of 469 trees. *Uvariopsis congensis* was most commonly used and accounted for 55.6 percent of all nests, yet its occurrence in Struhsaker's enumeration of trees is only 7.7 percent. Obviously *U. congensis* was a favorite nest species out of proportion to its availability. In addition it was a clumped species found in groves of low species diversity. Nests were often so clustered as to suggest a chimpanzee bedroom.

The upshot of these data is that chimpanzees tended to use particular species of common trees for their nests. The pattern of utilization of species apparently was influenced by the tensile suitability of the tree's limbs and its density of foliage and also by the proximity to a highly favored food source. I frequently observed a cluster of chimpanzee nests form around a large fruiting tree (e.g., *F. mucuso*) with additional nests appearing nightly during the period of fruiting.

Because of the nonrandom distribution of nest tree species and food species, nests were not randomly distributed. For a nest count to be representative it must result from a large enough transect. My experience suggests that 5.0 km is a minimum length; whereas 10 km is probably more desirable. T. Struhsaker (pers. comm.) pointed out that nest counts could best be interpreted by stratifying them according to habitat, such that a correction factor could be computed to interpret a density estimate for the entire study area. However, reliable identification and stratification of subhabitats within the forest habitat for that purpose was beyond the scope of this project.

Estimated heights of nests averaged 12.2 m at Ngogo ($R = 2-35$ m) and 10.8 m at Kanyawara ($R = 5-23$ m). The difference between means may be due primarily to the greater diversity of tall trees at Ngogo. I found no evidence that chimpanzees had nested on the ground. The average height of nests was 16.5 m near Fort Portal and in the Ruwenzori Mountains (Bolwig 1959), 9.1–12.2 m at Gombe (Goodall 1965), 19 m in the Kasakati Basin (Izawa and Itani 1966), 10 m in Rio Muni (Jones and Sabater Pi 1971), 11.7 m in Western French Guinea, and 11 m in

Kayonza (Schaller 1963). Variation in these means may reflect differences in vegetative physiognomies between habitats.

Estimation of chimpanzee density. Table 9 summarizes the computations of densities of chimpanzees in Kibale using data from nest counts along five transect routes. In the very few cases during secondary or tertiary counts where an old nest was found from a previous count, I did not count it again. Density estimates from the same route during different seasons may reflect seasonal use of the region by the nomadic chimpanzees. The highest densities are from compartment 30 in Kanyawara, $\bar{x}$ = 3.48 chimpanzees/km^2, and average 46 percent higher than those from Ngogo (computed from all counts in Kibale Nature Reserve), $\bar{x}$ = 2.38 chimpanzees/km^2. Struhsaker's (1975) 1970–1972 censuses at Kanyawara indicate a density of 1.4 chimpanzees/km^2, and his long-term record of chimpanzee vocalizations do not indicate an increase in density since then (pers. comm.). Because of these data I suspect that the high densities from my nest counts in compartment 30 may result from the chimpanzees' tendency to forage in peripheral, disturbed habitats (e.g., compartment 14) then move into compartment 30 before dusk to nest. Nesting sites are more scarce in the peripheral regions, except to the south, but food is present seasonally. I have observed members of the Kanyawara community moving 1–2 km from compartment 14 after feeding to compartment 30 during the afternoon.

The estimated 0.20 chimpanzees/km^2 along the census route through disturbed compartments 15, 14, 12, 13, and 17, a mosaic of areas partially felled and/or arboricided, may be an overestimate biased by the proximity of the south end of the route to Kanyawara. All three nests occurred within the first 3 of 9.5 km. Neither Struhsaker nor I saw chimpanzees along the route, although J. Skorupa (pers. comm.) has had rare glimpses of them feeding in emergent *Ficus* trees surrounded by secondary vegetation.

The habitat between compartment 30 and Ngogo is a mosaic of primary forest, colonizing thicket, and grassland. A trail cuts through these but only the section in primary forest was used for counting nests. The estimate of 2.04 chimpanzees/km^2 is based on a single short count.

The southern boundary of Kibale Nature Reserve is similar in aspect to the Ngogo study area, but may receive more pressure from poachers. It appears to be seasonal habitat for chimpanzees. The mean density for

TABLE 9. ESTIMATED DENSITIES OF POPULATIONS OF CHIMPANZEES IN KIBALE FOREST, UGANDA.

Enumeration route (north to south)	Date	(a) Nests counted in strip 20 m wide	(b) True length of route (km)[a]	(c) Area sampled =(b)x 0.02 km	(d) $\frac{(a)}{(c)(.8)}$	×	(e) $\frac{1}{\text{weighted mean life span of nests (days)}}$	×	(f) $\frac{1}{\frac{\text{Chimpanzee}}{\text{nest/day}}}$	×	(g) Ratio of whole population to nest-builders	=	(h) $\frac{\text{Chimpanzees}}{\text{km}^2}$	±	(i) 95% confidence interval
Compartment 15, 14, 13, 12 & 17	3 Mar. & 13 May 1978	3	9.53	0.191	19.6	×	$\frac{1}{119.2}$	×	(f)	×	$\frac{16}{13}$	=	0.20	±	
Kanyawara K30	6–11 Dec. 1977	55	10.09	0.202	340.3	×	$\frac{1}{100.1}$	×	(f)	×	$\frac{16}{13}$	=	4.18	±	0.945
Kanyawara K30	11–12 May 1978	45	9.37	0.187	300.8	×	$\frac{1}{102.4}$	×	(f)	×	$\frac{16}{13}$	=	3.61	±	0.945
Kanyawara K30	23–24 Mar. 1981	37	9.03	0.181	255.5	×	$\frac{1}{117.2}$	×	(f)	×	$\frac{24}{20}$	=	2.61	±	0.945
Kanyawara to N. boundary of Kibale N. R.	17 April 1981	15	5.50	0.110	170.5	×	$\frac{1}{100.1}$	×	(f)	×	$\frac{24}{20}$	=	2.04	±	0.945

Ngogo	6 Jan.–27 Mar. 1977	223	58.23	1.165	239.3	×	$\frac{1}{112.0}$	×	(f)	×	$\frac{46}{38}$	=	2.59	± 0.959
Ngogo	14 Jan.–17 Feb. 1978	35	9.68	0.194	225.5	×	$\frac{1}{101.3}$	×	(f)	×	$\frac{46}{38}$	=	2.69	± 0.959
Ngogo	18–20 Mar. 1981	30	10.00	0.200	187.5	×	$\frac{1}{115.3}$	×	(f)	×	$\frac{46}{38}$	=	1.97	± 0.959
S. boundary Kibale N. R.	15 Jan. 1978	1	5.50	0.110	11.4	×	$\frac{1}{107.7}$	×	(f)	×	$\frac{46}{38}$	=	0.13[b]	± 0.959
S. boundary Kibale N. R.	21 Mar. 1981	21	5.50	0.110	238.6	×	$\frac{1}{107.7}$	×	(f)	×	$\frac{46}{38}$	=	2.68	± 0.959
TOTAL for Kibale N. R.		310	88.91	1.778	217.9	×	$\frac{1}{111.0}$	×	(f)	×	$\frac{46}{38}$	=	2.38	± 0.959
TOTAL for all undisturbed habitats		462	122.90	2.458	234.9	×	$\frac{1}{109.4}$	×	(f)	×	1.21	=	2.60	

NOTE: Estimates are based on multiple enumerations of night nests of chimpanzees.
[a] Lengths of trail sampled twice, e.g., intersections, were reduced to a single length.
[b] During this count T. Struhsaker saw 3 chimpanzees. Several nests also were counted beyond the sample strip.

both my counts there, $\bar{x}$ = 1.4 chimpanzees/km^2, is probably skewed downward by my first count, when only one nest occurred within the sample strip while a dozen occurred just beyond it (see table 9).

The composite mean density (item 11 [h] of table 9) for all nest counts within undisturbed habitats in Kibale Nature Reserve is 2.38 chimpanzees/km^2 (± 0.959, 95% confidence interval). This figure probably can be applied to most areas of primary forest in the central block of Kibale Forest Reserve, but, as mentioned above, additional counts are to be desired before accepting this extrapolation uncritically (see also section below comparing censuses with nest counts).

Kano (1972) used a nest enumeration technique to estimate the density of chimpanzees in the Ugalla region of Tanzania, 50 km east of Lake Tanganyika between Gombe National Park and Mahale Mountains. Assuming a 6-month life span for nests, Kano concluded that Ugalla held 0.08 chimpanzees/km^2, an extremely diffuse population. Possibly a large portion of the area sampled was not appropriate habitat for chimpanzees but was included in Kano's analysis as if it were.

THE LINE-TRANSECT TECHNIQUE COMPARED WITH THE NEST-ENUMERATION TECHNIQUE

The estimate of population density from line-transect censuses, 1.45 chimpanzees/km^2 and the mean estimate from nest counts in Kibale Nature Reserve, 2.38 chimpanzees/km^2, are on a similar order of magnitude but are sufficiently different to warrant explanation.

The primary nest enumeration during 1977, including 65.5 percent of the transect length of the nature reserve sample, was counted during a season when chimpanzees were spending more time than usual in the study area (see tables 1 and 9). This probably skewed the entire count upward. In contrast, the line-transect censuses were conservative because I recorded only those apes I actually saw. Chimpanzees are cryptic, wary, easily frightened, and often quiet as they travel. They can be missed even when close. The census-based estimate of 1.45 chimpanzees/km^2 contains an inherent assumption that I was able to see and count every ape within the sample distance, which probably was not the case. Had both estimation techniques been executed flawlessly they might have yielded estimations more alike. For the further purposes of this study, the working estimate of chimpanzee population density is 1.45–2.38 chimpanzees/km^2.

The difference between my Kanyawara nest-count–based estimate of chimpanzee density, $\bar{x} = 3.11$ chimpanzees/km^2 (± 0.945, 95% confidence interval), and the census estimate of Struhsaker (1975) seven years earlier is discussed above. In the disturbed areas adjacent to Kanyawara—compartments 12, 13, 14, 15, and 17—a close correlation exists between my nest-count–based estimate and Struhsaker's (1975) and J. Skorupa's (pers. comm.) census-based estimates of chimpanzee density. All tended toward zero.

Rijksen (1978) reported that nest-count–based estimates of the population density of Sumatran orangutans were smaller by a factor of 15 than estimates based on apes actually seen. He concluded that such counts were too inaccurate to be trustworthy indicators of density. But Rijksen's transect lengths for nest counts averaged only 2.9 km ($R = 1.12$–5.35 km, $N = 11$), less than a third as long as what I would consider a desirable length for a transect of that purpose. Clearly further research on the efficacy and accuracy of both techniques of estimating chimpanzee population density would be useful.

The principal value of the nest-enumeration technique is that it requires a small investment of time, especially if the average life span of nests is estimated (3–6 months) rather than monitored for accuracy. Such monitoring may be precluded during population surveys of short duration or over wide geographical areas. A single transect route 10 km long usually requires two days of slow walking and careful scanning to locate and count nests with a strip 20 m wide. Multiple random transect routes are preferable to a single one. And recounts along the same routes at intervals of five months provide some indication of seasonal use. Seasonal use is also indicated at the age structure of the nest population. Observer efficiency in locating nests may be self-tested by recounting a transect in reverse. Where population structure is unknown, the data collected can still yield an estimate of density for the nest-building segment of the population.

On the other hand, multiple censuses along line transects of known length require regular and repeated visits to the sample area for a year or more. But such censuses may be more accurate than nest counts as indicators of ape density and they provide the following additional data: age-sex structure and condition of the population, feeding preferences and possible explanations for seasonality of use of the region, social structure, interspecific associations, and differential suitability of the habitat. Because some individuals may be missed within the sample strip, census estimates are conservative. Both methods should be used in regions to be surveyed thoroughly. Use of both may explain scarcity of chimpanzees in a region due to unsuitability of habitat, scarcity of nest sites, activities by human hunters, etc.

POPULATION DENSITIES OF CHIMPANZEES FROM OTHER STUDIES

Table 10 lists a series of estimates of densities of chimpanzees in several habitats in Central Africa. Only the estimates from Kibale Forest derived from systematic techniques incorporating line-transect censuses and nest counts. In general, the methods used to assess these densities were obscure; most were the impression of the workers.

A 32-fold range in estimates is evident. But perhaps significantly, the independent estimates from Kibale, Gombe, and Mahale Mountains (all relatively longer investigations) vary from 1.0–2.38 chimpanzees/km^2. Estimates from Budongo Forest might reflect seasonally high densities. Apparently, though, they were made post hoc without a systematic method of determining density in the field. Both considerations place the estimates by Reynolds and Reynolds (1965) and Sugiyama (1968) in a light difficult to interpret. Considering the remaining estimates in table 10 it appears that chimpanzees are more numerous in rain forests than in savanna-woodland habitats.

In contrast to this conclusion Sugiyama (1972:145) wrote, "Chimpanzees were long considered to be thick forest dwellers, but recent studies on their distribution and ecology reveal that population density is greater at the forest edge than in the interior of the high forest, and that they prefer the complex habitat of forest, woodland, and savanna (Suzuki 1969; Kano in press)." Note that Suzuki's (1969) estimate in table 10 for the density of chimpanzees in savanna woodland is 30–34 percent of the density I estimated for Ngogo while Kano's (1972:51) maximum estimate of mean density in woodland savanna was only 8.8–14.4 percent of that which I estimated for Ngogo. Surprisingly, Sugiyama himself (1968)

TABLE 10. POPULATION DENSITIES OF CHIMPANZEES IN CENTRAL AFRICA

Estimated density (chimpanzees/km^2)	Investigational technique	Area sampled (km^2)	Location of study	Source
1.45–1.95[a]	Line-transect censuses[b]	Unknown	Budongo Forest	Albrecht 1976:357
1.45–2.38	Line-transect censuses N=55 Nest counts along 88.9 km	~70	Kibale Nature Reserve	Ghiglieri, this study
1.29–1.93	Scouting; inference	77.7	Gombe National Park	Goodall 1968:168
0.21	Scouting; inference; Reports from locals; Nest counts	9,200	Malagarasi River to Karema Gap, (Tanzania)	Kano 1972:51
1.0	Scouting; inference	660	Mahale Mountains	Nishida 1968:213
2.90–3.90	Scouting; inference[b]	20.7	Budongo Forest	Reynolds and Reynolds 1965:393
1.4	Line-transect censuses N=44	<2	Kanyawara, Kibale Forest	Struhsaker 1975:290
6.7	Scouting; inference[b]	5	Budongo Forest	Sugiyama 1968:243
4.0–5.0	Head count in restricted habitat	5–6	Bossou, Guinea	Sugiyama and Koman 1979:329
0.49–0.71	Scouting; inference	747	Kasakati Basin, (Tanzania)	Suzuki 1969:131

[a] Density figure extrapolated from author's data on frequencies of vocalizations.
[b] Field investigations were less than one year in duration.

published the highest density for chimpanzees anywhere, 32 times greater than that of Kano (1972), for chimpanzees from his study in Budongo, a rain forest!

RANGING

Individual ranging

The ideal technique for gathering data on ranging is to follow a completely habituated individual all day, from night nest to night nest, and record the itinerary followed. Before research by expatriates at Gombe was truncated this technique was successfully employed by several workers (e.g., Wrangham 1975; Riss and Busse 1977; Pierce 1978; and see also MacKinnon 1974). My original intent at Ngogo naïvely included all-day follows of chimpanzees. But because only a few Ngogo chimpanzees gained a sufficient tolerance of me to make such follows feasible, and because of the limited visibility in the understory, the apes usually lost me during my attempts to follow. Beyond the trail-grid system the situation was even more difficult.

Data on daily ranging are limited to part-day follows of chimpanzees lasting 3.67–9.45 hours ($N = 12$). Nine part-day follows of a young adult female, Owl, covered a total of 62.4 hours and indicated a rate of travel ≥ 65 m/hour ($R = 100 - 1200$ m/follow). These data are conservative because I interpolated a straight-line itinerary between points of observation for periods when Owl was out of view. During more than 8 hours of continuous observation an old habituated male, Satan, traveled only about 100 m. A young male, Raw Patch, traveled ≥ 600 m during 6 hours. I followed a party of sixteen chimpanzees at Kanyawara as they traveled 500 m in 8½ hours. Another party of fourteen Kanyawaran chimpanzees traveled 1.5–2.0 km during 3½ hours of 7¼ hours of observation. On the following day two males of the same party of fourteen returned to their starting point of the previous morning. Because none of my follows were nest to nest, full day ranges for Kibale chimpanzees are still unknown.

The average day range of chimpanzees at Gombe was 3.9 km (Wrangham 1975); adult males exhibited significantly longer ranges than adult females, both on a daily and yearly basis. Pierce (1978) reported an average range of 2.6 km/day for members of the Kahama community at

Gombe. Clark (1977) reported that females with infants did not cover as great a range as did males and nulliparous females.

Estimation of community home range area

The nature of social structure among chimpanzees was one of their last attributes to clarify for researchers. Early reports from field studies did not hint at the existence of a permanent and discrete multimale social group (see chapters 6 and 7). Recent data from this and other studies, however, lead to the conclusion that males live in a xenophobic community and maintain the territorial integrity of their home range. Additional data from Pusey (1979) suggest that adult females may be equally involved in the community's spatial and social integrity. The home range occupied by a community is circumscribed by social and/or geographical boundaries recognized by its occupants (Nishida 1979).

From the topography and vegetation peripheral to Ngogo and from my follows of chimpanzees out of the study area, I suspect that the home range of the Ngogo community extended southward through the forest to and beyond the southern boundary of Kibale Nature Reserve toward the cultivated regions near Bigodi and also extended westward beyond Kansara grasslands to the Dura River and main road (see figure 2). Censuses, nest counts and vocalizations indicate chimpanzees were in these regions and the habitat was primary forest type contiguous with Ngogo. Complete data allowing the precise delineation of the home range of the community at Ngogo are lacking for reasons discussed in the section above. But it is possible to estimate the size of the home range given a few assumptions. The first is that the forest habitat of the entire region mentioned above is of equivalent carrying capacity over time despite seasonal fluctuations in the availability of food in localized portions of it. In other words the ecological value per unit of area of each subregion of the habitat is equal to that of the rest of the home range over the lifetimes of the apes inhabiting it. This assumption is untested because quantitative ecological data pertaining to the regions contiguous to Ngogo are lacking. My impression of these areas is that they are similar, though they likely contain some species of trees in different proportions and some different species of trees, any of which may exhibit different phenological cycles. The second assumption is that my estimates of population density of chimpanzees ($1.45-2.38$ apes/km^2) represent the range of densities for the entire home range of the Ngogo community. The third and most vulnerable assumption is that the Ngogo community contained about 55 members, a figure based on sightings of recognizable and unrecognized individuals during my first 17 months of research. Given that these assumptions are valid, the home range can be estimated by dividing community size by population density:

$$\frac{55 \text{ chimpanzees}}{\text{community}} \div \frac{1.45\text{--}2.38 \text{ chimpanzees}}{\text{km}^2} = \frac{23.1\text{--}37.9 \text{ km}^2}{\text{community}}$$

The community concept does not assume that every individual in the community actually traversed the $23.1\text{--}37.9$ km^2 during their lifetimes. The home range is habitually used by the interacting social network of 55 chimpanzees, some of whom may travel in only a portion of it.

An alternate means of estimating the home range area is provided by Milton and May (1976) who adapted McNab's (1963) general work on body size, food type and home range of birds and mammals specifically to primates. Milton and May found that the home range of an individual was related to individual biomass and diet. For primate hunters (frugivores and omnivores) the general relationship of body weight, diet and home range is:

$$\text{Log } HR_i = 0.83 \text{ Log } BW - 2.17$$

where HR_i is the home range (ha) of the individual and BW is body weight (g). Based on the community structure at Ngogo discussed above and on weights of free-living chimpanzees reported by Wrangham (1975) I estimated the weight of the average at Ngogo to be 28.7 kg (see also table 16 in chapter 5). According to the above formula from Milton and May (1976) the average home range portion per Ngogo chimpanzee is 0.34 km^2. For the community of 55 members the cumulative home range should be 18.6 km^2, an estimate somewhat smaller than that derived from my census and nest count data. The formula by Milton and May (1976) inherently underestimates the home range of chimpanzees because habitat type is not taken into account and because diet has been generalized to that of a hypothetical modal omnivore rather than a frugivore with more specific and demanding foraging requirements.

Wrangham (1975) computed the annual home range of an adult male chimpanzee, Figan, at Gombe to be 12 km^2 and equated that with the community range shared by eight other adult males. Pierce (1978) reported the same community (Kasakela) as ranging over 15 km² and also reported the neighboring Kahama community of five adult males as ranging over only 3.9 km². Soon after these data were recorded males from the Kasakela community annihilated the Kahama adult males (Goodall et al. 1979). Pierce (1978) suspected that the small home range of Kahama males was a product of intercommunity competition. A nulliparous female of the Kahama community ranged over 14.5 km² during the same year the males were restricted to 3.9 km². Nishida and Kawanaka (1972) estimated the home range of M-group, containing 60–80 chimpanzees, to be 17–≥34 km² in Mahale Mountains and the home range of the smaller, neighboring K-group, containing 24 members, to be 10.5–21.0 km². Some overlap occurred between the home ranges of the two communities.

POPULATION DYNAMICS

Births

Of the 19 adult females I identified in Kibale, 12 carried infants I judged to be 4 years old or less. Two Ngogo females, Kella and Mom, gave birth to infants, Kirk and Munch, during the first 17 months of this study. The period of these births is my estimate based on the sizes of the infants. Of the seven females without infants, two were accompanied by young juvenile offspring, three were mature or nearly so in appearance and exhibited sexual swellings, and two were very old, never had sexual swellings, and probably were post-menopausal. Graham and McClure (1977) reported a 48-year old captive female chimpanzee who continued cycling sexually but produced no offspring, and a similar 38-year old female who continued cycling but did not become pregnant due to ovarian senescence. Apparently, though, females in the wild do experience an apparent menopause and cessation of their estrous cycles (Pusey 1979; this study).

The chimpanzee community at Gombe contained approximately the same number of reproductive females as that at Ngogo (Riss and Busse 1977). Records of births (Goodall 1977) resulted in an average of 1.75 infants being born who survived beyond 1 month. Another 1.17 known or suspected births per year occurred in which the infants did not survive 1 month. Many of these early mortalities were known or suspected cases of intracommunity infanticide and cannibalism (Goodall 1977, 1979; and see Fossey 1978). I saw no evidence of similar events at Ngogo, but the birth rates of surviving infants seemed similar for both areas.

Mortality factors

All age-sex classes were symmetrically and well represented in Kibale with the possible exception of subadults at Ngogo. The population appeared to be healthy with little sign of mortality specific to a single age or sex class. At least two Ngogo chimpanzees disappeared due to unknown causes during the first 17 months of this study, a very old female, Hump, and an infant male 3 or 4 years old, Anson.

Crowned hawk-eagles (*Stepanoaëtus coronatus*) commonly soared above Kibale Forest and preyed on redtail monkeys (personal observation), black and white colobus (T. Struhsaker, pers. comm.), and gray-cheeked mangabeys (Freeland 1977). This massive forest eagle preys mainly on diurnal primates (Grossman and Hamlet 1964:308) and would be capable of carrying an infant chimpanzee away from the protection of its mother. But I never saw such an attack, possibly because most chimpanzee mothers maintained a close proximity to their infants. Other potential predators include spotted hyena and possibly leopards.

But within both Kibale communities, whose total membership may not have exceeded 100 apes, were 10 to 13 known or suspected victims of poachers' snares normally intended for ungulates. At Ngogo an adult male, Stump, and a young adult female, Ita, each had a healed over stump from which a hand had been amputated. At Kanyawara, most accessible to, and apparently more exploited by poachers, two adult males had healed hands missing some or all fingers, Hook and Ikarus; two adult males had a crippled mutilated foot, Klubfoot and a male reported to me by T. Struhsaker (pers. comm.); Struhsaker also saw an unidentified adult female missing a hand; I found an unnamed juvenile whose gangrenous foot had a snare wire embedded in it (she may have died later); another juvenile female, Donika, had a crippled hand with a snare wire embedded in her wrist; and T. Struhsaker was taken once by some local people from Kanyawara to free a juvenile male from a snare in which he had been caught.

Additionally, an informant told me that two or three chimpanzees, including an adolescent female and an adult male, who had been raiding sugar cane in Lwamugonera (6 km NNW of Kanyawara), were snared and speared during the final months of 1977 by professional poachers on contract to the cane growers. The extent of such mortality among the population in general is unknown.

Although Bakonjo reportedly killed and ate primates illegally in Kibale Forest prior to the 1962 civil war with the Batoro (S. M. Yongili, pers. comm. and heresay), the Batoro, current residents of the region around the central block of Kibale forest, consider primates unfit for human consumption. But, reportedly, when Batoro poachers found chimpanzees in a snare, they sometimes killed them for food for their dogs and because the skull can be sold for medicinal properties it is believed to possess and because captured infants may be kept or sold as pets. At present the intensity of deliberate and inadvertent human predation on chimpanzees (which is illegal in Uganda) is not known, but it may be a significant mortality factor near Kanyawara and possibly in other regions where the forest abuts cultivated regions.

Fossey (1981, 1982) reported similar predation and mutilations on the endangered mountain gorillas of *Parc National des Volcans* in Rwanda by local trappers. Only the most dedicated antipoaching efforts by workers such as Fossey and organizations such as the African Wildlife Leadership Foundation and conservationists such as Jean Pierre Von Vecke and Rwanda's Park Department forestall the extirpation of Rwanda's gorillas (Cahill 1981). Though currently not so endangered as Rwanda's gorillas, Uganda's chimpanzees may require similar attention in the future in the face of its human population growth.

4

FEEDING ECOLOGY AND PATCHINESS OF RESOURCES

The life history strategy of an animal is the result of natural selection having shaped a system in which resources are channeled into the production of offspring with an efficiency, or conversion rate, constrained by physiology and ecology (Gadgil and Bossert 1970; Schoener 1971). Gaulin (1979) pointed out that analysis of life histories usually considers two main components: reproductive strategy (morphological, physiological, and behavioral characteristics affecting resource conversion into the production of offspring) and feeding strategy (adaptations maximizing resource accrual). This chapter considers the feeding ecology of Kibale chimpanzees.

FEEDING ECOLOGY

An interplay exists between both components of resource conversion to offspring. Adult female chimpanzees give birth at five-year intervals (Goodall 1977; and see previous chapter). During any year most adult apes fail to produce offspring. This low birth rate, coupled with a long life span and prolonged social and environmental learning, makes chimpanzees a classic example of a K-selected species exhibiting a low r_m

(Cole 1954; Emlen 1973). In such a species, small increases in the efficiency of foraging may influence reproductive success.

As long as foraging is not efficient at the expense of reproductive activities, natural selection should favor a more efficient strategy (see Wrangham 1975). The foraging pattern of chimpanzees appears to be a highly adapted behavioral complex that *constrains* the sociality of the individual, though they remain highly social despite the constraints. Because almost any reproductive success among chimpanzees may be significant, subtle differences in the patterns of energy allocation and socialization may produce significant differences in reproductive success.

Activity patterns

Because of differences in reproductive problems faced by each sex, we would expect sexual differences in activity patterns. A female does not compete for a mate but does need to forage and range so that her dependent offspring receive adequate nutrition. Males do compete for mates and may defend and monopolize estrous females (Tutin 1975), but they do not need to range in accordance with the nutritional needs of their offspring.

Figure 6 shows the daily activity patterns of male and female chimpanzees beyond infancy in Kibale Forest. Data from all observations were tabulated for the categories foraging, resting, and traveling. Foraging includes picking, manipulating, eating, chewing, or carrying food items, and scanning for or examining food items by vision, touch, or smell. Resting includes all activities not included in foraging and traveling— e.g., most social interactions, nonforaging self-maintenance activities (grooming, defecating, etc.), play, sleep, or staring off into space. Traveling includes locomotor activities that took an animal from one general location in the forest to another. Displacements of a few meters within a single tree during foraging or resting were not considered traveling.

Males as a class compared to females as a class spent more overall time foraging (62.1% vs. 52.4%), less time resting (25.8% vs. 37.6%), and more time traveling (12.1% vs. 10.0%). These sexual differences in activity budgets were significant (Wilcoxon rank sum test $Te = 120, Tr = 180, n = 12, p \leq 0.05$; using data from 0700 to 1900 hours). Both sexes exhibited morning and late afternoon peaks in foraging activity separated by a midday peak in resting. Among males, though, foraging was

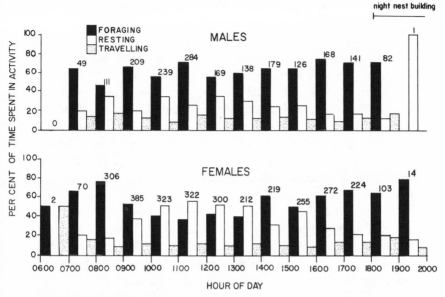

FIGURE 6. Diurnal activity patterns of male and female chimpanzees beyond infant age in Kibale Forest, Uganda. Data are summations of on-the-5-min observations during each hour. The sample period occurred between December 1976–May 1978. Number at the top of each column is sample size of observations for that hour.

the predominant activity during all hours of the day, whereas females rested more than half their time between 1000 and 1400 hours. Goodall (1968) reported only one daily feeding peak, between 1500 and 1800 hours, among Gombe chimpanzees.

My data are biased because approximately half of my observations were made during fruit-tree vigils. This tended to undersample traveling time. When considering only nonvigil data, though, traveling time was oversampled to a greater degree. The present combined data on males, however, differ little from a similar activity budget reported by Wrangham (1975). He found that adult males at Gombe spent 55.7 percent of their daylight hours foraging, 30.3 resting, and 13.8 traveling. Although they are approximately midway between the activity budgets I recorded for each sex, Wrangham's data on adult males more closely resemble my data on males than my data on females. Predictably, the time spent traveling reported by Wrangham was 1.7 percent higher than my data. Wrangham (1975) did not report female patterns.

The salient aspect of the activity budgets of Kibale chimpanzees is the significant difference between males and females—in particular, that males fed and traveled more than did females. Among other great apes, Rodman (1979) found that an adult male orangutan spent more time feeding and less time traveling than each of three females in his study area in East Kalimantan. MacKinnon (1974), however, reported no differences in the activity budgets of male and female orangutans in Sabah; perhaps because he included subadult males (who are of female size) in his analysis of male activity. Adult male orangutans are 2.2 times as heavy as adult females (Rodman, in press). Because of the metabolic demand of increased body weight (Brody 1945), males require more food for self-maintenance. Adult male chimpanzees are only 1.3 times as heavy as females (Rodman, in press), so the difference in body weights and associated metabolic demands is small. The metabolic demands of females, although smaller in weight, may equal or exceed that of males because of the demands of pregnancy, lactation, and the transport of an infant (see Gunther 1971), although the basal rates of males are somewhat higher than females by weight (Brody 1945) and their metabolic demands are increased by increased travel (Taylor et al. 1970). I suspect that the increased time spent foraging by males results from the increased metabolic demands of travel. Their increased travel may result from nonforaging concerns such as the need to patrol their communal home range and to increase their opportunities to locate and mate with estrous females. These sexual differences were predicted by the hypothesis of sexual selection and territoriality outlined in the introduction; their interpretation here assumes that males at Ngogo behave similarly to those in Gombe (Wrangham 1975; Goodall et al. 1979; and see chapters 6 and 7).

Food species and types

Plant foods. Chimpanzees in Kibale used 50 plant food types while I observed them (table 11) and were primarily frugivorous; 78 percent of feeding time was devoted to eating fruit pulp or pulp and seeds (instances of seed-eating only were not recorded as fruit consumption). Seeds of most fruits (such as *Ficus* spp.) were ingested with and as part of the fruit, either because to separate the seeds would greatly increase the handling time for each fruit or because seeds are nutritious. Digestion of seeds was not determined. Seeds of other species, such as *Monodora myristica* and *Mimusops bagshawei,* passed through the digestive tract of chimpanzees to emerge apparently undamaged in the feces. Young leaves eaten tended to be from tree species, such as *Chaetacme aristata* and *Celtis durandii,* whose leaves are unusually low in phenolics (McKey et al. 1978), a common chemical defense of plants against herbivory.

Reynolds and Reynolds (1965) reported Budongo chimpanzees as being 90 percent frugivorous. Hladik (1977) reported that provisioned chimpanzees at Ipassa, Gabon ate 68 percent fruit by weight in their nonprovisioned diet. Jones and Sabater Pi (1971) wrote that chimpanzees were "mostly frugivorous" in Rio Muni. Of the 80 food types of Gombe chimpanzees reported by Goodall (1968) 60 percent were fruit. Wrangham (1975) reported that adult males at Gombe spent 59.5 percent of their foraging time eating fruit. Izawa and Itani (1966) reported that fruit constituted 35.7 percent of food types of chimpanzees in the woodland savanna of Kasakati Basin. Although dietary composition reflects an interplay of preference and availability, these data suggest that a rain forest habitat provides a richer, more stable source of fruit than more open habitats (see Gaulin 1979:16).

Other plant food types eaten by Kibale chimpanzees included seeds, blossoms, bark, young and mature leaves, and leaf buds. My impression was that chimpanzees preferred fruit in most situations when they had a choice, but making any absolute statement concerning food preferences in the wild is not possible without knowing total availability of food and the bioenergetic costs associated with exploitation of each type (Emlen 1966; MacArthur and Pianka 1966; Schoener 1971). The few analyses of wild fruits which have been made (Janzen 1979) indicate that fruits of *Ficus,* the most common generic food of chimpanzees, are relatively high in protein and low in toxins when ripe. Such foods are a superior food type that, once located, can be harvested and processed quickly as discrete packets.

Harvesting of fruits appeared to be systematic and included much scanning of the crown. Rarely did an adult rework a section of tree crown that had been picked over in its presence by another adult. A slight tendency toward niche separation resulted from the ability of juveniles and

TABLE 11. SUMMARY OF OPPORTUNISTIC OBSERVATIONS OF CHIMPANZEES FEEDING.

Food species	Food type eaten								
	Fruit[a]	Seed[a]	Blossom	Bark	Cambium	Wood	Leaf bud	Young leaf	Mature leaf
Aphania senegalensis								2	
Celtis durandii	2	2					9	19	
Celtis mildbraedii (Budongo Forest only)								2	
Chaetacme aristata								3	
Cordia millenii	28								
Cynometra alexandri (Budongo Forest only)		2	13						
Ficus brachylepis	13	13							
Ficus capensis	4	4							
Ficus cyathistipula	5	5							
Ficus dawei	25	25							
Ficus exasperata	21	21							
Ficus kitubalu	2	2							
Ficus mucuso	131	131						11	
Ficus natalensis	49	49			1				24
Markhamia platycalyx					2			1 (petioles)	

Food species and type									
Mimusops bagshawei	19	19							
Monodora myristica	6	6							
Pseudospondias microcarpa	34	34						2	2
Pterygota mildbraedii	95[b]	37[b]	11	7		1			
Treculia africana		4	1						
Uvaria sp.								1	
Uvariopsis congensis	34								
Tree, unidentified								2	
Vine, unidentified								10	
Animal foods:									
Red colobus monkey, apparently, 1 observation									
Termites, unidentified, 2 observations									
Total: 51 types & species	468[a]	354[a] 6[c]	25	9	3	1	9	55	24
Percent of all 600 observations	78.0[a]	1.0[c]	4.2	1.5	0.5	0.2	1.5	9.2	4.0

NOTE: Data consist of numbers of on-the-five-minute interval observations during which a chimpanzee was feeding on a food species and type. Data are random with respect to observer expectations; data from fruit tree vigils are not included (see text). Data were collected in Kibale Forest (and Budongo Forest where noted), Uganda between December 1976–May 1978 and January–May 1981.

[a] During 74.4% of fruit eating (such as *Ficus* sp.) fruit and seeds were ingested together. Where such overlap occurred each feeding observation has been listed twice, but only the first figure was used to determine proportion of food type in diet. Digestion of seeds was undetermined.

[b] Food item eaten was wing of immature seed. Only one adult male was seen to ingest seeds.

[c] Seeds only eaten, not fruit.

small adult females to terminal-branch feed like a gibbon, *Hylobates lar* (Grand 1972), although a few large adults diminished this separation by bending a terminal branch in toward them to pick the fruits from it. The chimpanzees often tested figs for ripeness, without picking them. They did so by visual inspection and/or squeezing them gently. This conservative method of testing contrasted strikingly with wastage by several species of monkeys. I frequently observed gray-cheeked mangabeys and red colobus picking whole panicles containing as many as 30 fruits, but eating only part of one fruit and then dropping the rest.

Animal foods. Meat-eating and hunting of mammals and birds by chimpanzees have been reported by several workers from different study sites: Budongo Forest (Suzuki 1971), Gombe National Park (Goodall 1968; Teleki 1972; Wrangham 1975; Morris and Goodall 1977; Busse 1977, 1978), and Mahale Mountains (Kawabe 1966). During this study I saw one adult male, Eskimo, nibbling meat from the inner side of what appeared to be the skin of a red colobus. Eskimo carried the skin with him on two successive days, parts of which he spent foraging in a fruit-laden *F. mucuso*. I found no other parts of the carcass and do not know whether Eskimo made the kill or not. I never witnessed Kibale chimpanzees successfully capturing mammalian prey, although I did observe interactions between chimpanzees and monkeys that suggested hunting behavior and prey response (see section on interspecific interactions in chapter 5).

Chimpanzees have been observed to eat insect food more often than meat of mammals or birds (Azuma and Toyoshima 1962; Reynolds and Reynolds 1965; Goodall 1968; Nishida 1973; Wrangham 1975; Hamilton and Busse 1978). McGrew (1979) reported that adult female chimpanzees at Gombe foraged longer on insect foods than did adult males but significantly less on meat from mammals. Large termite nests built as conspicuous mounds, which are seen over much of sub-Saharan Africa, were virtually absent from my study areas in Kibale Forest. I observed chimpanzees eating unidentified termites in dead wood of a fallen tree. They dug the insects out of the rotting wood using the nail of a forefinger to excavate a hole up to 7 cm deep. Termites were eaten off the tip of the finger. I saw no use of tools for "termite fishing" or other harvesting of foods such as was reported by Merfield and Miller (1956), Goodall (1968), Struhsaker and Hunkeler (1971), Nishida (1973), and McGrew (1977). Neither did I observe a situation in which the use of such tools would have been advantageous at Ngogo. Perhaps necessity *is* the mother of invention.

Drinking water. Ngogo is divided by many small, perennial creeks that join the south-flowing Kanyanchu stream in the center of the study area. Chimpanzees drank at creeks by crouching low on all fours, placing their lips against the water and sucking it up. Knot holes in trees often contained water after rains but I saw only infant chimpanzees drink from these.

While feeding on figs some adults urinated copiously for 6–38 sec at intervals of 20–60 min. During 2.05 hours of observation of an old adult female, Gray, feeding on the figs of a *mucuso*, she urinated three times for a total of 91 seconds. Possibly many of the fruits eaten by chimpanzees contain sufficient water to obviate the need for drinking. Infants, who consumed little or no fruit, sometimes drank water in trees. One infant repeatedly rubbed the back of her hand on rain-soaked leaves then licked the water off her hand. Another infant excavated a hollow knot hole in a tree bole by scooping out an accumulation of detritus (onto his mother's lap). He repeatedly dipped his arm into the hole, then held it vertical until droplets streamed down toward his elbow. He licked them off his hairs before they passed his elbow. Goodall (1968) reported drinking water as common among Gombe chimpanzees, who used techniques similar to those described above, and also constructed leaf sponges to gain access to water in deep holes.

Spacing of foragers in fruit trees

Aggregations of up to 24 chimpanzees collected in large fruiting trees. Individuals beyond infancy consistently maintained an appreciable interindividual distance during foraging ($\bar{x} = 9.45$ m, $N = 1111$ 5-min observations). These separations often collapsed markedly after a period of intensive foraging when the animals formed small grooming clusters. As illustrated in figure 7, distances between nearest neighbors (excluding infants) during foraging decreased significantly with increasing size of feeding aggregations (correlation coefficient, $r = 0.6714$, d.f. $= 9$, $p \le 0.05$). Conversely, distances between foraging neighbors increased with increasing crown volume of the food tree (correlation coefficient, $r = 0.8173$, d.f. $= 8$, $p \le 0.01$). These data suggest that chimpanzees forage further away from a foraging conspecific when conditions allow them to

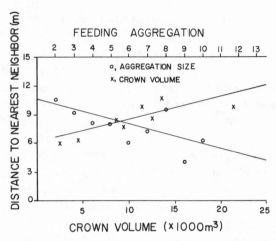

FIGURE 7. Mean distances between chimpanzee neighbors beyond infant age during foraging in relation to (1) size of feeding aggregation and (2) crown volume of food tree. Data were collected in Kibale Forest, Uganda between December 1976–May 1978.

reduce proximity without hindering foraging. This tendency to disperse, rather than maintain or increase proximity during foraging, was not only statistical, it was also my strong impression during observation. Individuals appeared to monitor one another in the crown and to move such that they did not increase their proximities. This pattern of behavior may contribute to the low number of dominance interactions I observed (chapter 6) despite a potentially competitive situation and implies an individual adaptation facilitating the chimpanzees' fusion-fission social system.

Food sharing

Various levels of food sharing among social mammals have been reported for a variety of taxa; wild dogs (*Lycaon pictus*) by Kuhme (1965), Estes and Goddard (1967) and Kruuk and Turner (1967); Serengeti lions by Schaller (1972); wolves (*Canis lupus*) by Murie (1944) and Mech (1970); and chimpanzees by Goodall (1968), Sugiyama (1972), Teleki (1972), McGrew (1975) and Silk (1978). Among chimpanzees the sharing of animal or vegetable foods was a solicited event; food was not shared by a donor unless asked for by a recipient via gestural or vocal communication (e.g., Fouts and Budd 1979; Silk pers. comm.). At Gombe unsolicited sharing of food by chimpanzees was virtually nonexistent (Silk *ibid*).

I observed two adult female chimpanzees sharing food at Ngogo: one instance was solicited, as is apparent from my field notes:

12 Nov. 1977, 1140, center 1470 *F*, 20m up in *Ficus mucuso*. AF (Clovis) is sitting feeding with IM (1.5) (Chita) in her lap. The infant pulls the female's right arm in toward him—she is holding a fig she just picked—and he eats part of the fig from her hand. The infant lets go of her arm, drops fragments of fig. Her mother holds the fig in the same position for another 10 sec or so, right where he had pulled it. He squirms around and shows no more interest in it. She pulls it up and eats it.

The second female who shared food apparently did so spontaneously and unsolicited:

20 April 1978, 1723, 10mW 1210 *C.5*, 19 m up in *F. dawei*. As AF (Mom) is climbing, scanning and picking fruit, she pauses, picks a fig, and hands it to her ventral-clinging infant (1–1.5, Munch) by putting the fig in the infant's mouth. She picks a second fig, puts it her own mouth and chews as she moves on. As she climbs upward I can see pieces of fig dropping from her infant's mouth as it chews. Note: Infant has remained ventral during all of my observation time. I did not see it beg gesturally nor vocally.

Mom shared figs with Munch repeatedly and intermittantly, apparently as a matter of course rather than in response to solicitation by her infant.

A mother sharing a fig with an unweaned infant is providing it with food which substitutes in part for milk she otherwise would provide during suckling. Because the metabolic conversion of food that the mother eats into milk is less than 100 percent efficient, such food sharing is an advantage if the infant can process the food item and otherwise is receiving adequate nutrition. Additionally, such sharing helps communicate to the infant which plants are edible and may promote autonomy, which would reduce parent–offspring conflict (Trivers 1974) during weaning. Silk (1978) reported that adult female chimpanzees at Gombe differentially responded to the solicitations of their infants, more often sharing with them food items that were difficult for the infants to procure than foods the infants could obtain easily.

Kano (1980) observed 261 instances of plant food sharing among pygmy chimpanzees (*Pan paniscus*) in Wamba. All age classes were recipients, infants were not donors, and sharing was solicited by the recipient and reluctantly shared by the donor. Large, rare food types, e.g., *Treculia africana,* were shared among adults. Provisioned sugar cane (*Saccarum officinale*) was the food shared most often. But foods difficult to process were shared most often with immature individuals.

In addition to the more classic cases of food sharing described above, male chimpanzees at Ngogo sometimes gave loud food calls advertising the presence of large food patches (see chapter 6).

PATCHINESS

Although Kibale chimpanzees paused while traveling to harvest foods en route, such as young leaves of a vine or understory tree, my impression was their basic foraging pattern involved intensive use of isolated patches of fruit.

Patch size and utilization

Patches—individual trees or groves of trees in ripe fruit—are irregularly distributed in the forest. Theoretically their size and density impose bioenergetic limits on the number of chimpanzees that can exploit them efficiently; hence, patches limit sociality to an extent. A patch containing only enough to provide three adults with acceptable food for an hour of foraging can provide only 6 min of foraging for the 30 adults of the Ngogo community. In a habitat of homogeneous value (assuming a model similar to that of MacArthur and Pianka 1966) the community traveling as a unit would need to travel 10 times farther than a party of three adults in order to visit a sufficient number of patches to feed everyone. A tenfold increase in traveling distance would increase each ape's metabolic requirements; furthermore, that much traveling would monopolize the daylight hours and leave no time for sedentary feeding or resting.

Unlike most primates living in social groups, chimpanzees seem always to have the option of leaving a party and traveling alone or with a smaller party. This open society, with its characteristically protean membership in subgroups (chapter 6), allows individuals to avoid competition with other community members when food resources are clumped and scarce by simply striking off on their own to forage solitarily on patches that might be uneconomical bioenergetically for a larger party to exploit.

Table 12 is a summary of my time spent during fruit tree vigils and of sizes of chimpanzee aggregations seen in 35 trees of varying crown volumes (Figure 11 illustrates dispersion of four species of vigil trees). I did not see chimpanzees feed in *F. congensis* and *F. polita* during vigils or other times, although redtail monkeys and red colobus ate from the former. I watched a juvenile male chimpanzee, Phantom, pass arboreally through the crown of a *F. polita* laden with apparently ripe fruit. He was traveling from one grove to another of *Uvariopsis congensis* and did not even test the figs. Chimpanzees reportedly ate figs of *F. Polita* at Gombe (Goodall 1968; Wrangham 1975) and at Budongo and the figs of *F. congensis*, also at Budongo (Reynolds and Reynolds 1965). Another vigil species in which I never observed chimpanzees during vigils was *F. brachylepis*, although I did observe chimpanzees eating the figs on other occasions.

The average size of all 687 feeding aggregations was 3.6 chimpanzees. Maximum aggregation sizes were significantly correlated with crown volumes of food trees (correlation coefficient, $r = 0.5653$, d.f. $= 18, p \leqslant 0.01$). Figure 8 shows the relationship between crown volume of vigil

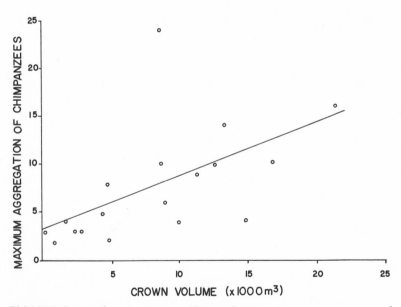

FIGURE 8. Patch size (N = 18) in relation to maximum sizes of chimpanzee aggregations (N = 687) in Kibale Nature Reserve, Uganda. See text for assessment of patch size.

TABLE 12. SUMMARY OF OBSERVER TIME SPENT DURING VIGILS OF FRUITING TREES AT NGOGO AND PRESENCE OF CHIMPANZEES PER TREE SPECIES.

Species of tree	No. of trees observed	Range in crown volumes (m³)	No. of days on vigil	Hours of vigil (hr:min)	Total chimpanzee hours in tree per vigil hour	Aggregation sizes of chimpanzees		
						No. of aggregations	Range in aggregation sizes	X̄ no. per aggregation
Cordia millenii	3	262–2,916	11	62:37	0.13	11	1–3	2.00
Ficus brachylepis	2	1,141–3,619	4	18:20	0	0	0	0
Ficus congensis	1	1,703	1	7:45	0	0	0	0
Ficus cyathistipula	1	884	9	42:49	0.06	5	1–2	1.20
Ficus dawei	4	8,183–13,261	49	478:14	0.81	279	1–14	3.23
Ficus exasperata	5	2,425–4,427	14	72:46	0.66	21	1–5	2.38
Ficus mucuso	14	4,781–21,303	51	393:10	1.06	352	1–24	4.12
Ficus natalensis	3	1,703–4,601	13	54:27	0.63	19	1–8	2.42
Ficus polita	2	3,470–4,965	3	17:05	0	0	0	0
Total 9 species	34	262–21,303	155	1047:13	0.78	687	1–24	3.60

NOTE: Data were collected between January 1977 and May 1978

trees and the maximum aggregations of chimpanzees I saw in each. Comparisons of average sizes of aggregations for all agggregations seen per tree (correlation coefficient, $r = 0.2191$, d.f. $= 18$, $p > 0.05$) and of average chimpanzee-hours per hour of vigil with patch size (correlation coefficient, $r = 0.2516$, d.f. $= 18$, $p > 0.05$), however, did not indicate significant differences, although I suspect a trend exists. Larger crown volumes seemed to support more feeders over time than did smaller crowns, but a bias in my sampling technique may have masked this trend. I usually spent an extra day or two toward the end of the fruiting period of a tree of large crown volume than I spent at a small tree. This differential cutoff for durations of vigils emphasized the peak fruiting period of smaller trees and the peak plus the dwindling fruiting period of larger trees, so that the utilization data per hour were not representative of the differences in crown volume over time. The following example illustrates the trends I suspect.

At Ngogo, *F. dawei* often developed into a very large dominant emergent from a parasitic epiphyte (Eggeling and Dale 1951). When in fruit the large trees attracted many chimpanzees. Parties arrived, fed, mingled, and sometimes remained in a single tree to rest or groom for as much as 6 to 9 hours. Patterns of utilization by different parties varied between large and small-crowned trees. Figures 9 and 10 represent patterns of visitation to a medium-sized *F. exasperata* (crown volume $= 4427$ m^3) between 3–10 July 1977 and to a large-sized *F. dawei* (crown volume $= 13{,}261$ m^3) between 16–26 April 1978. These two vigils are used as examples because of my relatively complete observation of their fruiting phases and because the influence of my presence was minimal at each. The crown volume of the *F. dawei* was approximately 3 times greater than that of the *F. exasperata,* as were the respective maximum aggregations of chimpanzees who fed in each (14 compared to 5) during the vigils. Utilization per unit time was also greater in the *F. dawei* (1.35 chimpanzee-hours/vigil hour) than in the *F. exasperata* (0.84 chimpanzee-hours/vigil hour). The number of different individuals who visited the *F. dawei* ($N = 36$) was over twice the number who visited the *F. exasperata* ($N = 17$). And the number of repeat visits by individuals to the larger tree was almost three times greater (1.11 repeat visits/vigil hour) than in the smaller tree (0.39 repeat visits/vigil hour).

Analysis of data from all vigils does not indicate significant differences

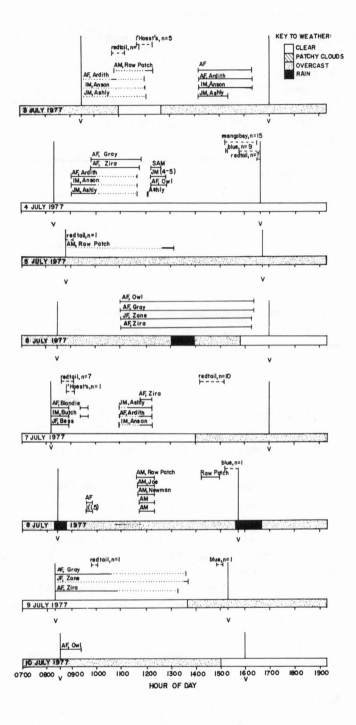

in utilization of different species of tree by chimpanzees if crown sizes are not considered, although such analysis is imperfect because of species differences in average crown sizes (correlation coefficient, $r = 0.4454$, d.f. $= 6$, $p > 0.05$), so crown volume per se may be the critical difference between the $F.$ $dawei$ and $F.$ $exasperata$ vigils discussed above. The upshot of these data from vigils is that larger crown volumes of fruiting trees attract more individuals, attract the same individuals for more repeat visits, sustain larger feeding aggregations, and are utilized more intensively over time than trees with smaller crown volumes. Because of the tendency of chimpanzees to aggregate in them, the density and distribution of large-crowned fruit trees may be an important factor in limiting and/or facilitating socialization between community members.

Table 13 summarizes sexual differences in patterns of visitation to $F.$ $dawei$ #4 during 16–26 April 1978, as illustrated in figure 10. On average, individual males visited on fewer days per individual, and visited fewer times per individual than did females. However, for each day that individual males visited $F.$ $dawei$ #4 they revisited the tree on the same day more often on average than did females. These sexual differences are not significant statistically (Wilcoxon 2 sample rank test, $n = 12$, $n_2 = 16$, $T_1 = 17.45$, $p > 0.05$) and may be a random result of sampling. But the tendency for males to be in a specific location for fewer days out of 11 than females on an individual basis and for those same males to revisit the superabundant food source more often than females on each day that they did visit is predicted by the hypothesis of sexual selection and territoriality outlined in the introduction. If males need to range over a greater area than females, then such ranging will result in males spending less time in any one area than do the more sedentary females, hence fewer days of visitation by the males. The metabolic cost of traveling from a distant area to return to forage at $F.$ $dawei$ #4 may have been more expensive than visits to local, though probably less abundant food

---

FIGURE 9. Eight-day record of visits to a fruiting Ficus exasperata by Ngogo chimpanzees and monkeys. Vertical lines with subscript, V, designate period of vigil per day. Horizontal dotted lines indicate that the chimpanzee was resting near the vigil tree. Dashed line indicates the presence of monkeys in the vigil tree. Lower horizontal bar indicates weather conditions for each day.

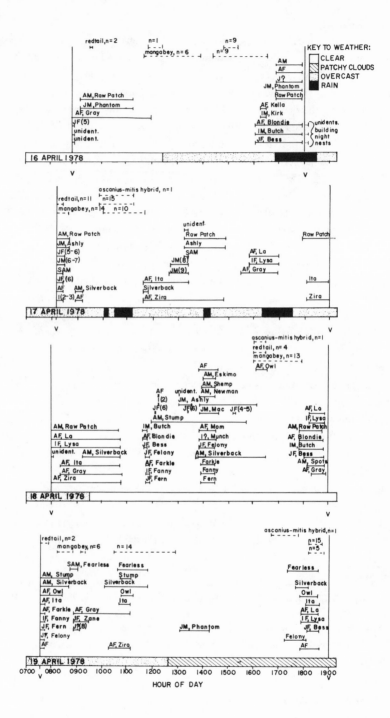

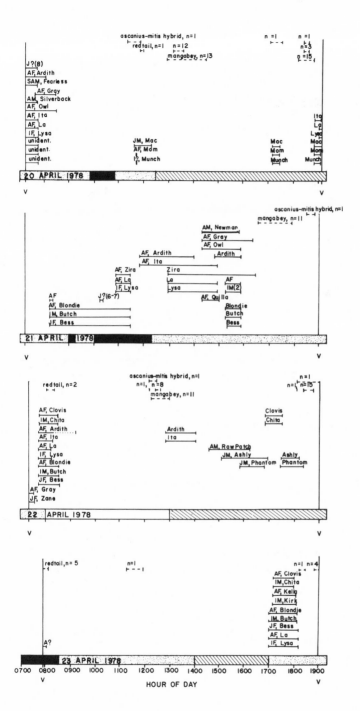

ascanius-mitis hybrid, n=1
redtail, n=1   n=12
mangabey, n=13

n=1   n=1
n=3
n=15

J?(8)
AF, Ardith
SAM, Fearless
AF, Gray
AM, Silverback
AF, Owl
AF, Ita
AF, La
IF, Lysa
unident.
unident.
unident.

JM, Mac
AF, Mdm
I?, Munch

Mac
Mom
Munch

Ita
La
Lysa
Mac
Mom
Munch

20 APRIL 1978

ascanius-mitis hybrid, n=1
mangabey, n=11

AM, Newman
AF, Gray
AF, Owl
AF, Ardith   Ardith
AF, Ita
AF, Zira   Zira
AF, La   La   AF
IF, Lysa   Lysa   IM(2)
AF   J?(6-7)   AF, Quilla
AF, Blondie   Blondie
IM, Butch   Butch
JF, Bess   Bess

21 APRIL 1978

ascanius-mitis hybrid, n=1
n=1, n=8
mangabey, n=11

redtail, n=2

n=1
n=1   n=15

AF, Clovis   Clovis
IM, Chita   Chita
AF, Ardith
AF, Ita   Ardith
AF, La   Ita
IF, Lysa   AM, Raw Patch
AF, Blondie   JM, Ashly   Ashly
IM, Butch   JM, Phantom   Phantom
JF, Bess
AF, Gray
JF, Zane

22 APRIL 1978

redtail, n= 5   n=1

n=1 n=4
AF, Clovis
IM, Chita
AF, Kela
IM, Kirk
AF, Blondie
IM, Butch
JF, Bess
AF, La
A?   IF, Lysa

23 APRIL 1978

0700 0800 0900 1000 1100 1200 1300 1400 1500 1600 1700 1800 1900

HOUR OF DAY

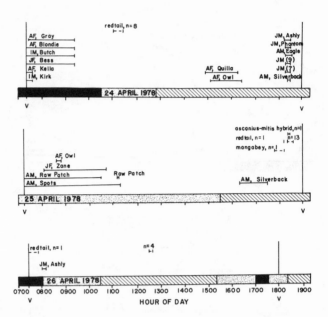

FIGURE 10. Eleven-day record of visits to a fruiting *Ficus dawei* by Ngogo chimpanzees and monkeys. Vertical lines with subscript, V, designate period of vigil per day. Horizontal dotted lines indicate that the chimpanzee was resting near the vigil tree. Dashed line indicates the presence of monkeys in the vigil tree. Lower horizontal bar indicates weather conditions for each day.

sources. The greater number of return visits by males when they were in the vicinity of *F. dawei* #4 may have resulted from their greater metabolic demands generated by larger body size and/or more time spent traveling in general. The average duration of visits (not truncated by my presence) by males ($\bar{x} = 74$ min, $R = 1$–251 min, $N = 43$) to *F. dawei* #4 was less than, but similar to, visits by females ($\bar{x} = 79$ min, $R = 1$–210 min, $N = 70$).

Patch density and distribution

Patch density. The Ngogo study area contains approximately 100 species of trees, yet relatively few were important food species of chimpan-

TABLE 13. SUMMARY OF VISITS BY INDIVIDUALLY KNOWN, POST-INFANT-AGED CHIMPANZEES TO A FRUITING *FICUS DAWEI* (PHENOLOGY SPECIMEN #4) BETWEEN 16–26 APRIL 1978 AT NGOGO.

Sex of visitor to tree	Total no. of visitor days	Range in visitor days per individual chimpanzee	S.D.	X̄ no. of visitor days per individual chimpanzee	Total visits	Range in visits per individual chimpanzee	S.D.	X̄ no. of visits per individual chimpanzee	X̄ no. visits per day per days visited by individual chimpanzee
Female, N=16	62	2–8	2.03	3.88	86	2–10	2.60	5.38	1.39
Male, N=12	34	1–6	1.90	2.83	53	1–11	3.55	4.42	1.56

NOTE: See also text and figure 10.

zees. Using T. Struhsaker's unpublished tree enumeration data for Ngogo, plus my enumeration of *Ficus* spp. (table 14), I examined the relationship between relative density of food species and their importance as food. Compared with the average density of all trees ≥ 10 m tall enumerated by T. Struhsaker, *Ficus* spp., for example, include only 0.52 percent by number of all tree per ha at Ngogo. Table 15 compares the rank in relative density of the top 12 food species of Ngogo chimpanzees (as determined from opportunistic, nonvigil observations of feeding in table 11) with their rank of inclusion in diet. The difference between the two rank orders is significant (Sign test, $n+ = 10$, two-sided, $p \leq 0.05$).

This comparison illustrates an obvious but important aspect of foraging: chimpanzees are specialist feeders who preferentially exploit rare food *types* (see section on types above) of rare *species*. This tendency is in opposition to some of the more common primate species, such as red colobus, which feed on more common food types (e.g., leaves) of more common food species in Kibale (Struhsaker 1975).

The composition of chimpanzee diet is an unusual exception to the

TABLE 14. DENSITY OF *FICUS* SPECIES AT NGOGO, KIBALE FOREST, UGANDA.

Species	No. of trees counted	Density trees/ha
Ficus brachylepis	66	0.53
F. brachypoda	1	0.01
F. capensis	6	0.05
F. congensis	3	0.02
F. cyathistipula	22	0.18
F. dawei	24	0.19
F. exasperata	33	0.26
F. kitubalu (?)	3	0.02
F. mucuso	29	0.23
F. natalensis	37	0.30
F. polita	11	0.09
F. pseudomangifera	9	0.07
F. vallis-choudae	not sampled systematically	
Total	244	1.95

NOTE: Data were obtained by enumeration of all specimens with roots reaching the ground within a strip 20m x 62,375m, i.e. 124.75 ha.

TABLE 15. TOP 12 FORAGE SPECIES OF CHIMPANZEES COMPARED TO RELATIVE DENSITY AMONG TREE SPECIES AT NGOGO.

Tree species	Rank as food species	Rank in relative density (all trees)
Ficus mucuso	1	87[a]
Pterygota mildbraedii	2	8
F. natalensis	3	35
Pseudospondias microcarpa	4	20
Uvariopsis congensis	5	2
F. dawei	6	63
Cordia millenii	7	40
F. exasperata	8	45
Mimusops bagshawei	9	18
F. brachylepis	10	34
Monodora myristica	11	14
Celtis durandii	12	3

NOTE: Relative density was determined from unpublished data of T. Struhsaker (see text).
[a] F. mucuso did not appear among the 86 species enumerated by Struhsaker (but did appear in my enumeration of Ficus spp.) so I have arbitrarily assigned it the subsequent rank.

general mammalian trend of increasing body size and concomitant high absolute metabolic requirements linked with the exploitation of ever more common food resources. Drawing on the work on savanna herbivores observed by Bell (1971), Geist (1974), and Jarman (1974), Gaulin (1979:8) summed up this relationship nicely:

because of their high daily total food requirements, large animals will usually be unable to base their diets primarily on rare food items. They should eat those items that rank high in e/t [see Schoener 1971] whenever they are encountered, but low abundance relative to the animals' needs will prevent high-quality foods from comprising a large percentage of the diet. On the other hand, due to lower per-unit-weight food requirements, large animals do not need to provide a high rate of nutrient flow to their tissues, and are thus able to subsist on low-quality foods. For small animals the problems are reversed.

By specializing on relatively rare patches of fruit, chimpanzees have taken a narrow ecological path which requires superior abilities to locate and exploit such resources (see below). In addition, a large-bodied specialist's diet of rare food types dispersed in discrete patches implies the need for a large home range (McNab 1963) and for increased time spent traveling between patches (MacArthur and Pianka 1966).

Patch distribution. Existing data on the distributions of tree species, arising from enumerations along transects, have not been analyzed in terms of distributional tendencies. However, my observations in the forest and the mapping of enumerated trees both indicate a clear tendency among 9 of the 12 species in table 15 toward nonrandom and irregular distributions. Most species exhibited fidelities to microhabitats.

For the purpose of illustrating this, figure 11 depicts the sociability (*vide* Oosting 1956) of the four primary species of *Ficus* used by chimpanzees. *F. dawei* tended to be the most hyperdispersed; most specimens enumerated ($N = 45$) were in swamp forest, in the nadir of v-shaped creek courses, or in low points of relief where moving ground water was concentrated. Only nine specimens grew ≥ 100 m from a perennial watercourse. *F. exasperata* tended toward hyperdispersion along well-drained slopes; none were seen in swamp forest associations. *F. natalensis* was a more eurytopic species occurring in swamp forest and along well-drained slopes. *F. mucuso* tended to be a social species, clumping in very diffuse groves well dispersed from one another in deep forest on well-drained sites. Other *Ficus* species not used as intensively as the four above exhibited similar tendencies of hyperdispersion and fidelity to microhabitat. Edaphic factors and/or plant physiologies resulting in these nonrandom distributions are unknown.

Hubbell (1979:1304) found a similar trend among rare tree species in a dry tropical rain forest, "When total population size is considered, the trend is toward increasing clumping with decreasing abundance." Of 61 species Hubbell enumerated, 72 percent exhibited significant clumping among adults; the remaining species appeared randomly distributed. The mechanisms of plant sociability were not clear cut.

I gained an impression of the distributions of other unmapped food tree species during repeated walks in the forest. None seemed randomly distributed. *Cordia millenii*, *Mimusops bagshawei*, and *Pterygota mildbraedii* were primary forest emergents usually found on well-drained slopes. *Pseudospondias microcarpa* was a swamp forest species. *Ficus brachylepis* and *Monodora myristica* were eurytopic but rarely found in swamps. *Uvariopsis congensis* was a numerous and ubiquitous understory species of primary and secondary forest that often grew in dense groves containing only a few other species. The distributional trends of yet other food species used less intensively also showed varying habitat or associational

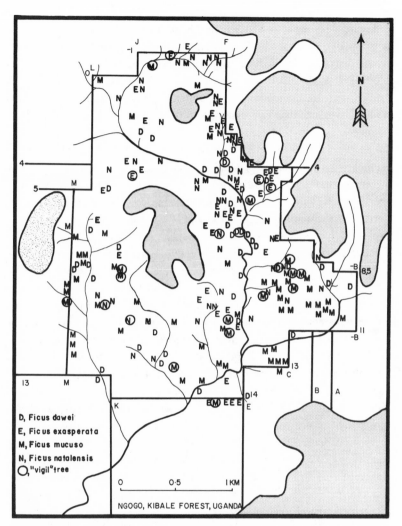

FIGURE 11. Distribution of the four major species of *Ficus* used as food resources by chimpanzees at Ngogo, Kibale Forest, Uganda. Note that this sample includes only specimens visible from the trail-grid system.

tendencies. Often it was possible to gauge the likelihood of finding one species by noting the presence or absence of other associated species.

Figure 11 shows that, from the perspective of four *Ficus* species at least, Ngogo was not a uniform habitat of any particular time with respect to the unavailability of food from any single species. Other species exhibiting a patchy distribution on a wide scale included *Pterygota mildbraedii*, rare in compartment 30, common in the center of the Ngogo trail-grid system, then rare again south of trail 10. Only one specimen of *F. mucuso* occurred at Kanyawara despite the numbers found at Ngogo. These two species were the top food species of Ngogo chimpanzees. Summarizing in advance of more data and statistical analysis, Ngogo is a complex mosaic of repeating floristic associations which, from the perspective of a chimpanzee, would be viewed as coarse-grained and patchy. Terrestrial locomotion seems an essential adaptation for chimpanzees to exploit such resources and still remain in social groups.

Patch phenology. A spatial assessment of patchiness with regard to the distribution of fruit eaten by chimpanzees makes little sense without considering the phenology of the patches. While the spatial distribution of food species changes very slowly during a chimpanzee's life, edible fruit on the trees appears and vanishes weekly. During the week or so that a tree bears a ripe crop, the greatest quantity of ripe fruit apparently is available in early morning and steadily becomes less so as a variety of diurnal primates and birds forage through it. This was suggested by further analysis of the visits by chimpanzees to *F. dawei* #4 presented in figure 10. Considering only those visits when premature exits were not caused by my presence, analysis indicates that visits made before noon ($\bar{x}$ = 1.65 hours, N = 36) lasted 47.3 percent longer than visits after noon ($\bar{x}$ = 1.12 hours, N = 77). This pattern seems a reflection of availability of ripe fruit. But more than this fine-grained diel component of temporal patchiness, the corase-grained seasonal aspect of fruit availability seemed critical to chimpanzees.

Figure 12 illustrates the fruiting periodicity of 12 species of food trees

---

FIGURE 12. Fruiting phenology of twelve top food species used by chimpanzees at Ngogo, Kibale Forest, Uganda. Number at head of column indicates number of trees with fruit. See text for explanation of phenological evaluation.

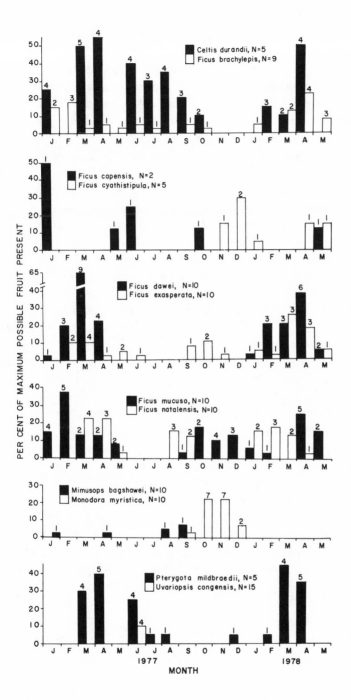

of Ngogo chimpanzees. A quick glance at it reveals no marked season for all species included, but, instead, a series of fruiting peaks for mixed species. Fruiting synchrony was most apparent in *F. dawei, M. myristica,* and *P. mildbraedii,* which fruited during peaks centered around March–April, November–December, and March–April respectively at 12-month intervals. Other *P. mildbraedii* specimens at Ngogo fruited out of phase with those monitored for phenology ($N = 5$) and were an important food of chimpanzees. Individual *F. mucuso* fruited at odd, unpredictable intervals ($\bar{x} = 6.55$ months, $R = 4 –13$ months, $N = 20$ fruiting periods) with fruiting synchrony possibly occurring by chance. *U. congensis* showed a marked fruiting peak during June 1977, but the 15 phenology specimens were relatively low producers. *F. brachylepis* and *F. natalensis* appeared to be asynchronous with intervals between fruiting periods varying from 5 to >17 months. Waser (1975) reported similar overall fruiting periodicity among several food species (including *Ficus* spp.) of mangabeys at Kanyawara.

Phenology from the above specimens indicates a general unpredictability for single species on a fine level. A similar apparent phenological anarchy was commented upon by Richards (1952:199), "In evergreen tropical forest flowering generally extends throughout the year and there is no season in which a proportion of the species are not in flower, some of them blossoming almost continuously; but even in the least seasonal climates there are maxima of flowering at certain times of the year, which are, however, often not clearly enough marked to strike a casual observer."

But by combining the data from Ngogo for all 101 phenology trees of species used by chimpanzees for food, for 17 months, a conspicuous pair of peaks in fruiting became evident from February–April of both 1977 and 1978. Figure 13 compares this composite fruiting periodicity (and availability) for fruit at Ngogo with the month to month record of my sightings of chimpanzees. The average number of chimpanzees I encountered per hour in the forest at Ngogo (table 1) was significantly and positively correlated with the percent of maximum possible fruit available on a monthly basis (correlation coefficient, $r = 0.6486$, d.f. $= 17$, $p \leqslant 0.01$). My impression, unfortunately not well documented, is that many of the Ngogo apes migrated south from the trail-grid system when the availability of fruit decreased within it.

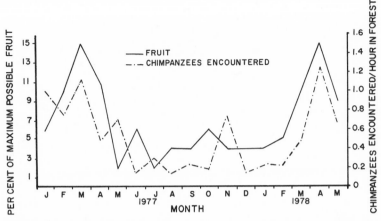

FIGURE 13. Chimpanzees seen per hour in the forest at Ngogo compared to the percentage of fruit present per month. Fruit presence is considered as a fraction of maximum possible fruit yield per total number (N = 101) of phenology food trees of chimpanzees in any one month. See text.

A visual comparison of average size of traveling parties per month with percentage of fruit present per month (table 16, figure 14) reveals a correlational trend not statistically significant (correlation coefficient, $r = 0.3374$, d.f. $= 17$, $p > 0.05$). I suspect that had I included a wider variety of food species (e.g., *Cordia millenii* and *Pseudospondias microcarpa*) in my phenology, the greater representation of fruit availability might have yielded a significant correlation with size of traveling parties. My impression was that the apes did travel in smaller parties during lean periods; such a response seems predictable on the basis of bioenergetic constraints associated with traveling longer distances to visit an adequate number of food patches (McNab 1963; MacArthur and Pianka 1966; Taylor et al. 1970). Smaller parties need to travel to fewer patches and hence are more efficient.

During my two years at Ngogo I came to think of the forest in phenological terms. I sometimes visualized it as if it were possible to record periodicity by exposing a hypothetical "fruit-sensitive" motion picture film for a year from an aerial perspective above Ngogo and then speed up the exposed film to produce a time-lapse effect. The image I saw was

TABLE 16. SUMMARY OF SIZES OF PARTIES OF CHIMPANZEES SEEN PER MONTH AT NGOGO, KIBALE FOREST, UGANDA.

Month	No. of parties seen	$\bar{X}$ no. of chimpanzees per party	Range in party size	S.D.
December 1976	1	1.00	N/A	N/A
January 1977	70	2.64	1–9	1.82
February 1977	30	3.83	1–24	4.75
March 1977	76	2.66	1–10	2.25
April 1977	26	2.74	1–8	1.88
May 1977	46	2.43	1–8	1.88
June 1977	14	1.64	1–3	0.84
July 1977	32	1.97	1–5	1.15
August 1977	10	1.60	1–3	0.84
September 1977	12	1.75	1–4	1.22
October 1977	1	3.00	N/A	N/A
November 1977	22	3.64	1–18	4.34
December 1977	8	1.13	1–2	0.35
January 1978	24	1.96	1–6	1.40
February 1978	11	1.73	1–3	0.79
March 1978	37	2.41	1–10	1.80
April 1978	89	2.57	1–12	2.13
May 1978	18	2.94	1–8	2.15
Total 18 months	527	2.56	1–24	

a series of tiny winking blotches of similar hue (i.e., the same species) appearing simultaneously, to be replaced quickly by blotches of a new hue and so on. Occasionally, isolated blotches of an odd hue would appear out of synchrony in the mist of other hues. At times winking blotches would dominate the scene, at other times they would appear as isolated phenomena in a drab field. For chimpanzees the greatest concentration of flickers would be times of plenty. During the lean times competition would be at its most severe and their social system would be under its greatest strain.

Because of the unusual ecological problems faced by chimpanzees as large-bodied social apes exploiting rare food types, no matter what the availability of fruit we would expect them to exhibit unusual abilities of resource location. Chimpanzees with the best developed spatial sense and memory, form sense (for recognizing plant species), and time sense (for

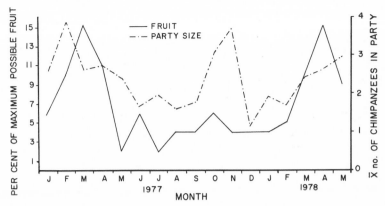

FIGURE 14. Average size of traveling parties of chimpanzees at Ngogo compared to the percentage of fruit present per month. Fruit presence is considered as a fraction of maximum possible fruit yield per total number (N = 101) of phenology food trees of chimpanzees in any one month. See text.

understanding the phenomenon of synchronous fruiting) will have a competitive advantage over other apes and sympatric monkeys. Studies of mental abilities of captive chimpanzees indicate that, when food is an incentive, they are capable of insightful learning (Kohler 1925), arithmetic learning (Ferster 1964), discrimination task learning (Robbins and Bush 1973), symbolic learning (Fouts and Budd 1979), efficient planning of a food-gathering itinerary (Menzel 1973), and communication of the locations of hidden foods between individuals (Menzel 1979).

Menzel (1973) conducted experiments in spatial memory by a delayed-response variation of the "traveling salesman" combinatorial problem. A juvenile chimpanzee was carried in a large outdoor enclosure and shown the positions of 18 randomly placed, hidden foods. Then the ape was isolated for 2 min and released to collect the food. The six subjects found an average of 12.5 pieces of food, scoring most highly on the 9 pieces of hidden fruit, and adopted an average itinerary 64 percent as long as the mean of all possible itineraries—an indication of selective spatial memory. Wolfe (1936) and Cowels (1937) demonstrated that chimpanzees would perform abstract discrimination tasks to collect as many as 30 food tokens of varying values "in anticipation" of purchasing a 30-token incentive.

It is not surprising that Wrangham (1975) admitted that Gombe chimpanzees were "good botanists" who appeared to search differentially in the correct plant community for species in season. They also searched for fruit in each specimen encountered of a synchronously fruiting species. These relatively advanced mental abilities are helpful, if not essential, adaptations allowing chimpanzees to successfully compete with thousands of sympatric anthropoid primates for scattered and ephemeral food patches irregularly distributed within the apes' home range.

5

INTERSPECIFIC INTERACTIONS

Proximity to sympatric anthropoid primates was an important aspect of the immediate environment of Kibale chimpanzees. The cost per chimpanzee of competition between individuals of a feeding aggregation may be outweighed before and after feeding by social benefits conferring increased inclusive and individual fitness. But the cost of competition with monkeys apparently is not offset by benefits associated with proximity per se, unless an ape has a greater opportunity to prey on a monkey because of their closer proximity during scramble competition for superabundant resources.

Chimpanzees have been observed to kill and eat monkeys (Goodall 1968; Suzuki 1971; Teleki 1972, Wrangham 1975; Busse 1977; Morris and Goodall 1977), which apparently are a highly preferred food type. Such kills were observed infrequently however, and probably are rare events within most populations studied. It is likely that some individuals, especially females, never capture a monkey and consume a full meal from it during their whole lives (see McGrew 1979).

More obvious was the apparent competition for fruit between monkeys and chimpanzees, who occasionally defended such resources from monkeys. Chimpanzees in Kibale also clashed with monkeys in apparent attempts to prey on them. Defensibility of fruit trees varied with their size and growth forms but seemed primarily determined by the representation of age-sex-reproductive classes and the size of the social groups of each species and also by the species of monkey in contest with chimpanzees. During such encounters red colobus, for instance, invariably supplanted chimpanzees, who invariably supplanted redtail monkeys. I had the impression that relations between the apes and monkeys were never neutral; competition and/or potential predation seemed to be an unrelenting influence.

Unfortunately, owing to limitations of time and personnel, it was not possible to conduct observations of the foraging and behavior of sympatric primates during the period of observations on chimpanzees. All observations of foraging by monkeys at Ngogo were opportunistic on my part and are supplemented by personal communications from T. Struhsaker and S. Wallis. As such they are incomplete and provide only an indicator of levels of competition between apes and monkeys. But the competitive interactions between them sometimes were striking: I do not think it is possible to understand chimpanzee ecology without considering competition with sympatric anthropoids.

BIOMASS DENSITY

The process of converting estimates of numbers to biomass in Kibale Forest follows the work of Struhsaker (1975:186, 292) for compartment 30. The present treatment shares one of the weak points of his analysis: a relative paucity of data on the weights of free-living primates. My use of Wrangham's (1975) weights for Gombe chimpanzees is the only improvement in this regard. Table 17 summarizes my estimates of biomass density in Kibale Nature Reserve for the mammal species listed in table 5. More than 85 percent of the total by primates consisted of three species: red colobus (37.9%), baboon (30.0%), and redtail monkeys (17.8%). In contrast, my census data indicated the chimpanzees accounted for only 2.9 percent of the total, being outmassed by the folivorous red colobus 13:1. Clearly chimpanzees are relatively rare primates even in undisturbed habitats where hunting pressure is minimal.

Struhsaker's (1975:292) estimate for total biomass density in compartment 30, 2,217.4 kg/km^2, is substantially higher than my estimate at Ngogo, only 10 km south. However, red colobus accounted for 79.4

TABLE 17. BIOMASS DENSITIES FOR ANTHROPOID PRIMATES AND UNGULATES IN KIBALE NATURE RESERVE, UGANDA.

Species	$\bar{X}$ group size[a]	$\bar{X}$ group composition and (weight in kg)[b]			$\bar{X}$ group biomass (kg)	$\bar{X}$ density per 100 ha		Total biomass per 100 ha (kg)
		Adult males	Adult females	Immatures		Groups	Solitaries or individuals	
Primates								
Chimpanzee	55	24.3% (41.9)	27.1% (34.3)	48.6% (19.0)	1,577.9	0.026	1.45	41.6
Baboon	45	21.0% (34.0)	23.5% (13.6)	55.5% (11.6)	745.9	0.57	0.21	432.3
Redtail monkey	35	2.9% (4.0)	34.3% (2.9)	62.8% (1.5)	71.8	3.51	0.77	255.1
Blue monkey	16	6.3% (6.0)	34.0% (3.0)	59.7% (1.5)	36.7	0.12	0.05	4.7
l'Hoest's monkey	10	10.0% (6.0)	30.0% (3.0)	60.0% (1.5)	24.0	0.05	0.10	1.8
Mangabey	15	20.0% (10.5)	46.7% (7.0)	33.3% (3.5)	98.0	1.28	0.15	127.0
Red colobus	50	17.0% (10.5)	30.0% (7.0)	53.0% (3.5)	287.0	1.90	0.05	545.8
Black & white colobus	9	11.1% (10.5)	33.4% (7.0)	55.5% (4.5)	49.0	0.54	0.15	28.0
					Subtotal, primate biomass = 1,436.3			
Ungulates								
Bushpig	3, 8	63%	(61.2)	37% (30.6)			0.52	25.9
Giant forest hog	1, 3		(181.4)	(90.7)			0.31	45.8
Red duiker	1		(11.3)	(5.7)			7.95	73.4
Blue duiker	1		(7.9)	(4.0)			1.86	12.0
Bushbuck	1		(54.4)	(27.2)			1.03	47.9
					Subtotal, ungulate biomass = 205.0			
					Total biomass = 1,641.3 kg			

NOTE: Data were obtained during 55 line-transect censuses

[a] References for group size: baboon (Rowell 1966); mangabey, redtail and red colobus (Struhsaker 1975, 1977, and pers. comm. resp.); black and white colobus (Oates 1974); others, this study.

[b] References for weights: chimpanzees (Wrangham 1975; Dorst & Dandelot 1970 for age-sex sizes); baboon and ungulates (Dorst & Dandelot 1970); all others (Struhsaker 1975).

percent of the compartment 30 total; baboons were not present at all. Aside from these striking differences, which are discussed in chapter 3, the remaining biomass estimates are similar. Additionally, Struhsaker's estimates of population densities near the southwestern boundary of Kibale Nature Reserve are similar to mine. Relative to the nature reserve, compartment 30 contains a superabundance of red colobus. It is not clear whether their high density is due to superiority of habitat or to immigration of monkeys from recently disturbed habitats adjacent to compartment 30. J. Skorupa (pers. comm.) suspects the former is the case.

From a general perspective of animal ecology, the above data from Ngogo provide the basis for comparing primary mammalian consumer biomass in a rainforest with such densities in other habitats. Ignoring the rare, but biologically important, presence of elephant, buffalo, waterbuck, rodents, and strictly nocturnal mammals (not included in censuses), my estimate in Kibale Nature Reserve is only 37.3 percent of the 4431 kg of nonvolant mammalian biomass/kg^2 of rainforest at Barro Colorado reported by Eisenberg and Thorington (1973). At Barro Colorado tree sloths (*Bradypus* and *Choloepus*) accounted for 62.7 percent of the biomass above, while primates accounted for only 12.0 percent. The mammalian biomass from Ngogo also is only 40.8 percent of the 4027 kg of ungulates/km^2 reported by Watson et al. (1969) for the Serengeti ecosystem, or 32.8 percent of the 5000 kg/km^2 reported by Hendrichs (1970) for the same area. I suspect that had the large herbivores of Kibale not been decimated by hunting (e.g., Eltringham and Malpas 1976), and had my censuses been appropriate for such fauna, the total biomass for Ngogo would have been substantially higher. Interpretation of these densities depends upon analyses of the palatable components of the primary productivities of the habitats surveyed.

A major value of biomass estimates is in formulating an index of interspecific competition based not only on overlap in diet but also on the relative biomass of the competing species.

INTERSPECIFIC COMPETITION

The Kibale chimpanzees were feeding specialists who concentrated their foraging on large trees bearing ripe fruit. Such resources attracted several other species competing directly with chimpanzees. Competition is used here in the sense of Emlen (1973) in that each species experiences a depressed fitness because of the presence of the other at a food source. This study revealed no unequivocal evidence that any population of primates was limited by sympatric foragers, but the design of the study was not addressed at such issues. All evidence of competition is proximal and opportunistic and serves only to indicate rather than quantitatively define. Competition in fruit trees could be argued as trivial if it is assumed that large trees hold enough for all foragers but, in fact, observations of individuals arriving after long feeding bouts by chimpanzees suggested that such bouts sometimes exhausted the supply of the ripest fruit from even large trees (see section on patch phenology in the previous chapter). More to the point, the active expulsion by chimpanzees or monkeys of other species foraging or about to forage in a tree suggests that fruit was limited and competition was occurring. One dimension of competition is dietary overlap.

Dietary overlap

Data collected during fruit tree vigils include a running log of presence of sympatric anthropoid primates and their time of foraging in association with chimpanzees. Table 18 summarizes the coexploitation of seven important food species by diurnal primates at Ngogo. One indicator of competition provided is the ratio of hours one or more monkeys spent in a vigil tree per hour of vigil. Mangabeys and redtail monkeys spent the greatest amount of time in such trees, 29.26 and 24.67 hours of which (respectively) was in association with one or more chimpanzees. In contrast, one or more red colobus spent 130 hours of my total vigil time in two species of vigil trees, but only 24 min of that time in association with the apes. During most of these 24 min the two species engaged in aggressive interactions.

A more representative indicator of feeding competition is the extent of dietary overlap for specific food types. Table 19 presents the percent overlap in food types between chimpanzees and monkeys based on non-systematic, unquantified, and opportunistic observations of monkeys.

Redtail monkeys fed upon nearly half the plant food types I observed chimpanzees eating. Redtails have been termed frugivore-insectivores at Ngogo (T. Struhsaker pers. comm.). Mangabeys overlapped the apes by 38 percent. Waser (1975) reported that mangabeys at Kanyawara were

TABLE 18. CO-EXPLOITATION OF SOME PRIMARY FOODS OF CHIMPANZEES BY SYMPATRIC ANTHROPOIDS DURING 1,022.38 HOURS OF TREE VIGILS BETWEEN JANUARY 1977 AND MAY 1978 AT NGOGO, KIBALE FOREST, UGANDA.

Tree species	Chimpanzee hours in tree per vigil hour	Monkey hours in tree/vigil hour and (hour:min ≥ 1 monkey in association with chimpanzee)							
		Baboon	Redtail monkey	Ascanius-mitis-hybrid	Blue monkey	l'Hoest's monkey	Mangabey	Red colobus	Black and white colobus
Cordia millenii	0.13	0.004 (0:11)	0	0	0	0	0	0	0
Ficus brachylepis	0[a]	0	0	0	0	0	0	0	0
Ficus cyathistipula	0.06	0	0.07 (0)	0	0.001 (0)	0	0.68 (0)	0	0
Ficus dawei	0.81	0.09 (1:50)	0.54 (7:32)	0.02 (2:30)	0.29 (2:48)	0	1.59 (15:39)	0.04 (0)	0
Ficus exasperata	0.66	0.02 (0)	0.21 (0:14)	0	0.10 (0)	0.003 (0:04)	0.27 (0)	0	0
Ficus mucuso	1.06	0.03 (0:31)	0.19 (16:38)	0	0.0001 (0:03)	0	0.43 (13:20)	0.30 (0:24)	0
Ficus natalensis	0.63	0	0.04 (0:17)	0	0	0	0.03 (0:17)	0	0
Total		0.04 (2:32)	0.29 (24:41)	0.007 (2:30)	0.11 (2:51)	0.0002 (0:04)	0.79 (29:16)	0.13 (0:24)	0

[a]Chimpanzees were seen to feed on the figs of Ficus brachylepis at Ngogo and Kanyawara but not during vigil hours.

59 percent frugivorous (considering time spent feeding). They may be the closest competitors of chimpanzees. Additional data on blue monkeys will almost certainly indicate an overlap greater than the 32 percent shown in table 19. At Kanyawara they were frugivore-generalists (Rudran 1976). Baboons and red colobus, leaders in biomass representation, also took 24 and 20 percent respectively of their food types from among those used by chimpanzees. I never saw black and white colobus at Ngogo feed on food types taken by the Ngogo apes, yet four of the top food types of black and white colobus at Kanyawara (Oates 1974) were chimpanzee foods at Ngogo. Pending availability of lists approximating the total dietary diversity and proportions of monkeys at Ngogo, table 19 is my best, yet less than satisfactory, indicator of dietary overlap with chimpanzees.

Other ubiquitous, but small competitors included red-legged sun squirrels (*Heliosciurus rufobrachium*) and a variety of birds. Family groups of turacos (*Corythaeola* and *Tauraco*) and Black and White Casqued hornbills (*Bycanistes*) took fruit from vigil trees. Chimpanzees and sometimes redtails actively chased them from fruit trees.

Polyspecific associations

Table 20 summarizes all systematic sightings during 57 censuses of primates in polyspecific and monospecific associations in Kibale Forest. The major difficulty in analyzing the species-specific propensity to be in either mode of association is determining the random probability of either mode. Without knowing the exact patterns of spatial utilization for all species present, the density for each species, and the average area each social group occupies over time, random probabilities for frequency of overlap of groups, i.e., polyspecific associations, cannot be generated. Such an analysis would be complicated by the tendency for conspecific groups to avoid one another and by patchiness, the differential suitability of portions of the habitat to fill the needs of foraging and protection from predation.

The Yates-corrected Chi-square values in table 20 analyze the relative tendency for species to be in polyspecific association. For this analysis the expected value of polyspecific association was generated from the ratio of all incidents of monospecific association versus all incidents of polyspecific association in the pooled sightings ($N = 316$) during censuses of *all* species. The pooled proportion multipled by the number of

TABLE 19. COMPETITORS FOR FOODS OF CHIMPANZEES AT NGOGO.

Food species	Anthropoid primates							Rodent	Birds						
	Baboon	Redtail monkey	Blue monkey	l'Hoest's monkey	Mangabey	Red colobus	Black & white colobus	Heliosciurus rufobrachium	Treron australis granviki	Tauraco schutii	Corythaeola cristata	Psittacus erithacus	Bycanistes subcylindricus	Gymnobucco bonapartei	Lamprocolus purpuriceps
Celtis africana		F,S													
Celtis durandii		F,S yl[a]	F		F,S		yl								
Chaetacme aristata				yl?		yl		F,S,							
Connopharyngia holstii						yl[a]									
Cordia millenii	F	F													
Ficus brachylepis	F	F,S	F,S		F,S	F,S		F,S			F,S				
Ficus capensis													F,S		

Species															
Ficus cyathistipula	F,S		F,S		F,S			F,S		F,S	F,S			F,S	F,S
Ficus dawei	F,S	F,S	F,S		F,S	F,S		F,S	F,S	F,S	F,S			F,S	F,S
Ficus exasperata	F,S	F,S	F,S	F,S	F,S	F,S		F,S	F,S	F,S	F,S			F,S	F,S
Ficus mucuso	F,S	F,S			F,S			F,S		F,S	F,S			F,S	
Ficus natalensis	F,S	F,S	F,S	F,S?	F,S	F,S		F,S	F,S	F,S	F,S			F,S	F,S
Mimusops bagshawei	F,[a]S,[a]		F,[b]S,[b]		F,S			F,S							
Markhamia platycalyx	F,S	F,S	F,S		F,S										
Monodora myristica	F,S	F,S	F,S		F,S			S							
Pseudospondias microcarpa						S		S		S	S	S			
Pterygota mildbraedii				Ws,ml[a]											
Treculia africana			S[a]		F			F		F					
Uvariopsis congensis	F	F	F		F			F		F	F				
TOTAL:	12	24	16	4	19	10	1	18	4	8	11	1	10	6	4
% overlap with 50 food types taken by chimpanzees	24	48	32	8	38	20	2	36	8	16	22	2	20	12	8

[a] Observed by Thomas T. Struhsaker, pers. comm.

[b] Observed by Simon Wallis, pers. comm.

NOTE: Symbols for food types are: B, blossom; bl, leaf bud; F, fruit; S, seed; yl, young leaf; ml, mature leaf; Wd, wood; Ws, wing of seed. Unless otherwise noted, foods of monkeys were opportunistic observations.

TABLE 20. TENDENCY FOR ANTHROPOID PRIMATES IN KIBALE FOREST, UGANDA TO FORM POLYSPECIFIC ASSOCIATIONS.

Species	Number of sightings: groups and (solitaries)	Frequency in association				Chi-square values (Yates-corrected)	Probability (P≤)	Normal mode of association
		Polyspecific		Monospecific				
		Observed	Expected	Observed	Expected			
Chimpanzee	16(parties)	2	6.5	14	9.5	6.478	0.025	monospecific
Baboon	16 (4)	3 (0)	6.5 (1.6)	13 (4)	9.5 (2.4)	4.145 *	0.05	monospecific
Redtail monkey	98 (17)	57 (1)	39.7 (6.9)	41 (16)	58.3 (10.1)	12.288 (9.991)	0.0005 (0.005)	polyspecific (monospecific)
Blue monkey	5 (1)	4 (0)	2 —	1 (1)	3 —	* *		
I'Hoest's monkey	1 (3)	1 (0)	— —	0 (3)	— —	* *		
Mangabey	38 (4)	19 (2)	15.4 (1.6)	19 (2)	22.3 (2.4)	0.976 *	0.35	probably polyspecific
Red colobus	88 (2)	50 (1)	35.6 —	38 (1)	52.4 —	9.114 *	0.005	polyspecific
Black & white colobus	20 (3)	6 (0)	8.1 —	14 (3)	11.9 —	1.403 *	0.25	probably monospecific

NOTE: Data were collected during 57 line-transect censuses. See text for discussion of probability of frequency of polyspecific associations. Open numbers = groups; solitaries = parentheses.
*Sample size too small for statistical analysis.

1. Kibale Nature Reserve and Ruwenzori glaciers in background (photo by C. S. Ghiglieri).

2. Motorable track through primary forest southeast of Kibale Nature Reserve (photo by C. S. Ghiglieri).

3. Ngogo camp and a portion of the study area extending southwest beyond it (photo by C. S. Ghiglieri).

4. A typical view across a narrow belt of swamp forest at Ngogo (photo by M. P. Ghiglieri).

5. Swamp vegetation along the southern boundary of Kibale Nature Reserve (photo by M. P. Ghiglieri).

6. Buttresses of *Ficus mucuso* specimen #1 at Ngogo. I estimated its crown volume to be 21,303 m³, the largest at Ngogo (photo by M. P. Ghiglieri).

7. Elephants in a partially felled region immediately north of Kany-awara (photo by J. P. Skorupa).

8. Red duiker (photo by J. P. Skorupa).

9. Medan Mukasa, one of the Ugandan game guards assigned to Ngogo, with spears and snares confiscated after chasing poachers at Ngogo (photo by M. P. Ghiglieri).

10. Red colobus foraging on leaves (photo by L. A. Isbell).

11. Baboons in the forest at Ngogo (photo by C. S. Ghiglieri).

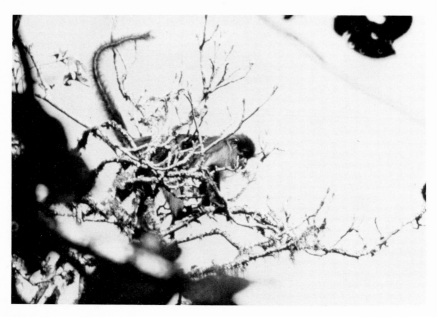

12. Gray-cheeked Mangabey at Ngogo (photo by M. P. Ghiglieri).

13. A redtail monkey leaping a gap in swamp forest at Ngogo (photo by M. P. Ghiglieri).

14. Solitary male blue monkey at Ngogo. Occasionally these usurp redtail harem masters and mate with female redtails to produce hybrids (photo by C. S. Ghiglieri).

15. Crowned Hawk-eagle in Kibale Forest Reserve, a predator on several species of primates (photo by C. S. Ghiglieri).

16. Talons of immature Crowned Hawk-eagle that apparently starved north of Kanyawara in Kibale Forest. The raptors apparently use these talons to pierce the skulls of their prey at first contact (photo by J. P. Skorupa).

17. A chimpanzee night nest (photo by M. P. Ghiglieri).

18. Chimpanzee in a day nest at Kanyawara (photo by J. P. Skorupa).

19. Adult male chimpanzee, Kong, at Kanyawara (photo by C. S. Ghiglieri).

20. Adult female chimpanzee, Gray, with a handful and mouthful of *Ficus mucuso* fig wadge at Ngogo. Very ripe figs were not wadged (photo by C. S. Ghiglieri).

21. Adult male chimpanzee, Eskimo, foraging in a *Ficus mucuso* at Ngogo (photo by C. S. Ghiglieri).

22. Subadult male chimpanzee, Fearless, and young adult female, Owl, foraging in a *Ficus mucuso* at Ngogo. Note size of Owl's perineum as sexual signal (photo by M. P. Ghiglieri).

23. Chimpanzees often rested during midday without building a day nest (photo by J. P. Skorupa).

sightings for each species yields an expected frequency for the occurrenc
of each associational class for that species. Because each value is only a
relative measure comparing the pattern of each species to the conglom-
erate pattern for all eight species, they provide a very conservative test.
For instance, if the tendencies of *all* species to form polyspecific associations were high but unequal, this test might indicate that the least gre-
garious species were not likely to be in polyspecific association at all.
Because this test was generated by the animals themselves, I found it
preferable to the arbitrary expected value of 50 percent used by Struhsaker (1975) for a similar analysis.

Chimpanzees ($p \leqslant 0.025$) and baboons ($p \leqslant 0.05$) were primarily
monospecific. In contrast, redtail monkeys ($p \leqslant 0.0005$) and red colobus
($p \leqslant 0.005$) were statistically likely to be in polyspecific association,
often with one another. Male redtails without a harem, however, were
likely to be solitary ($p \leqslant 0.0005$). As predicted above, social groups of
mangabeys cannot be described as significantly likely to be in polyspecific association even though they were so on half the sightings (usually
with redtails), a proportion which intuitively seems significant. Waser
(1980) reported that mangabeys at Kanyawara were in polyspecific association during 28 percent of all scans using a 20-m interspecific distance
criterion and during 56 percent of all scans using a 50-m distance criterion. Waser associated this high frequency of association with similarities
in the foraging needs between species and also discussed such associations as possibly enhancing evasion of predators. The results of my analysis above agree basically with those of Struhsaker (1975).

The ecology of polyspecific associations is incompletely understood.
Gautier and Gautier-Hion (1969) and Gautier-Hion and Gautier (1974)
attributed semipermanent associations of *Cercopithecus* spp. and *Cercocebus* sp. in Gabon to enhanced foraging success and to predator avoid-
ance, as did Waser (1980) above and Gartlan and Struhsaker (1972) for
associations of five species of *Cercopithecus, Mandrillus leucophaeus,
Cercocebus torquatus,* and *Colobus satanas.* Klein and Klein (1973) attributed nonrandom associations of *Cebus apella* and *Saimiri sciureus* to
enhanced foraging success for *C. apella.* See Waser (1980) for a more
detailed summary of the functions of polyspecific associations.

The forces favoring polyspecific associations among monkeys in Kibale apparently did not include chimpanzees. Gartlan and Struhsaker

(1972) reported also that chimpanzees were not included in the frequent associations of anthropoids in Cameroon. Despite their propensity for monospecific associations, at times Kibale chimpanzees did associate and interact with monkeys. Column e of table 22 (below) lists the percentages of my total observation time that Kibale chimpanzees spent in such associations. Almost all of this time was spent foraging in large fruit trees. In one exceptional and brief instance five species, chimpanzees, redtails, blue monkeys, mangabeys, and red colobus, were clustered in a foraging association at least 100 m in diameter. In general, unless groups of *Cercopithecus* spp., mangabeys or baboons contained very young infants, and the chimpanzee party contained one or more adult males, the monkeys did not forsake feeding in a superabundant food source. When in association monkeys usually did not allow close proximity to adult male chimpanzees or females not attended by immature offspring. In a few situations monkeys and/or apes increased mutual proximity in response to one another; all such observations are described in the section on interspecific interactions below.

Interspecific interactions

Although occasionally too subtle for me to discern, most interactions between chimpanzees and monkeys merely involved each species looking at the other, then either maintaining, reducing or increasing their proximity. These outcomes seemed to have been influenced by the size and age-class representations of each species' social group. Competition for food seemed to be the major aspect of most interactions. Table 21 summarizes dominance relationships between chimpanzees and sympatric anthropoid primates in Kibale. Sample sizes for most species are too small for Chi-square analysis; trends for others are unequivocal. The episodes which provided data for table 21 are enumerated and/or described below. I saw no overt interactions between chimpanzees and blue monkeys or l'Hoest's monkeys.

Baboons. Interactions between chimpanzees and baboons were usually peaceful. Feeding often occurred in close proximity. Once I observed an adult male chimpanzee, Silverback, and an adult male baboon traveling and arriving within 3 sec of one another at a *Cordia millenii* which went unvisited by other primates during the 9 hours of vigil that day. The two fed together until the baboon spotted me and left. On the other hand, an

TABLE 21. INTERSPECIFIC DOMINANCE INTERACTIONS BETWEEN CHIMPANZEES AND OTHER ANTHROPOID PRIMATE SPECIES AT NGOGO, KIBALE FOREST, UGANDA.

	No. of dominance interactions with chimpanzees					
	Baboon	Redtail monkey	ascanius-mitis hybrid	Mangabey	Red colobus	Black & white colobus
Chimpanzees supplanted other species	2 (100)	9 (100)	2 (100)	5 (71.4)	0	0
Other species supplanted chimpanzees	0	0	0	2 (28.6)	15 (100)	1 (100)
Total (36)	2	9	2	7	15	1

NOTE: (Quantity) represents percent of total interactions per pair of species.

old female chimpanzee, Gray, actively displaced a juvenile baboon by slapping her palm flat against a limb of a *F. mucuso*. A young female ape, Owl, approached a juvenile baboon in a *F. dawei*. The baboon glanced at her and quickly moved away. Chimpanzee relations with baboons in Gombe National Park ran the gamut from friendly play to competition to theft of kills to aggression to predation (Goodall 1971a, b; Teleki 1972; Morris and Goodall 1977).

Redtail monkeys. Chimpanzee relations with redtails were strictly unidirectional, dominance-subordinate interactions of three types: (1) passive dominance in which the redtails moved away in response to the mere presence of seemingly inattentive apes ($N = 3$); (2) ambiguous dominance in which redtails retreated from the approach or display of a chimpanzee when the redtails' presence seemed incidental to the chimpanzee but may not have been ($N = 4$); and (3) active supplantation by a chimpanzee who feinted toward or actively chased redtails away from a fruit tree. I never saw redtails make an aggressive move toward a chimpanzee of any age class.

ascanius-mitis hybrid. Because *ascanius-mitis* hybrids were rare, I saw only two interactions between them and chimpanzees. They followed the same patterns as (1) and (3) above between redtails and the apes. During the first a female hybrid rushed out of tree along with her group of redtails as an adult male chimpanzee approached. The hybrid involved in the second observation was without his harem of redtails and feeding about 5 m from an adult male chimpanzee, Silverback, in a *F. dawei*

19 April 1978. 1743. 10 m W 1210 *C.5*. S.B. moves toward the central portion of the crown from a terminal branch at a normal, leisurely pace. As he passes beneath the branch upon which the hybrid is foraging, S.B. suddenly grabs the limb and shakes it rapidly and vigorously. The hybrid leaps about 5 m to another limb then runs another 5 m away. S.B. does not pursue him. 5 min later the hybrid has left the tree.

Gray-cheeked mangabeys. All but one overt interaction occurred when mangabeys fed in association with the apes. Instances of passive dominance, as described above for redtails, were ambiguous with mangabeys because of their typically loose group cohesion (see Waser and Floody 1975). When only a few of several mangabeys departed a fruit tree in conjunction with the arrival of chimpanzees I could not confidently interpret their exit as a response to the apes. On two unequivocal occasions mangabeys deserted a tree in response to the arrival of apes. On another occasion an adult male mangabey approached a juvenile female chimpanzee, Bess, who, when she saw him, quickly moved 6 m away to sit and scan for fruit. The mangabey supplanted Bess and also scanned from her original position. On three occasions chimpanzees displayed and/or rushed at mangabeys in apparent attempts to maintain sole proprietorship of a fruit tree. Only one attempt, a weak one, failed:

19 April 1978, 1810, 10 m W 1210 *C.5*, As an adult male mangabey enters *F. dawei* #4 Owl (a young adult female), who is sitting above his point of entry, stomps on a dead limb to send it crashing downward past the mangabey. He enters the tree despite this but moves away from Owl.

The final and most complicated interaction between chimpanzees and mangabeys is best viewed through my notes:

17 April 1978, 0825, 15 m N *8.5*. Many monkeys (redtails and mangabeys) move S.S.W. out of *F. d.* #4 in a rushing wave, perhaps in response to chimps in that section of the crown [7 post-infant-aged chimps + 1 infant].

0837. Many redtails and mangabeys have moved S.S.W. and are lingering in the swamp. Between 0827–0829 6 chimps exit *F. d.* #4 in the same direction. As I look for R. P. [the only adult male chimpanzee present] in *F. d.* #4, redtails and mangabeys erupt into simultaneous, high-intensity alarm calls. I crane my head to see a large bird glide down the swamp corridor away from the monkeys. I guess it may have been an eagle, and the alarm calls were directed toward it. But the alarm calls suddenly intensify and I see redtails and mangabeys streaming away from a nearby tree that is vibrating as if a vigorous struggle were taking place in it.

0838. I hear a brief chimp scream. Two or more animals are scrambling through the undergrowth toward me. I move a few m closer along the trail. Galloping like a horse a large juvenile male chimp rushes toward me. Only 3–5 m behind him an adult male mangabey is chasing. The chimp appears not to see me only 4 m away, turns his head to view the mangabey, and swings with one arm around a sapling to perch on a horizontal limb about 0.6 m above the trail. The mangabey followed so closely that when the chimp stopped he found himself just below his feet. The chimp turns and they grapple for about 1 sec.

0839. Then the chimp leaps down running S. W. and the mangabey runs W. along *8.5* then climbs a tree. A moment of quiet ensues, then new alarm calls erupt, a rather small "large" bird glides nearby and a new scuffle begins in the original tree. This lasts only a few seconds. Amid the redtail and mangabey alarm calls I hear a single, high-pitched chimp scream sounding as if it were retreating from me. [I guess that the first chimp took advantage of the monkeys dropping low in response to the raptor to make an attempt to nab a monkey. Apparently at least one AM mangabey interrupted and the chimp ran. The incident may have been reinacted at 0839 but with different principals.] The mangabeys move to the edge of the swamp where they linger [all group members present]. The redtails move away further. [I do not know if all group members were present.]

Although attempted predation seems a likely explanation for the episode above, strictly speaking, it must be classified as a dominance interaction between a chimpanzee and a mangabey.

The interactions between the two species seemed to have been determined by several factors already mentioned, but the age-sex composition of the respective groups was one of the most important. For example, in March 1977, mangabey group S.B. at Ngogo spent approximately 2 hours foraging in a *F. mucuso* in association with two adult female chimpanzees. The attitude of the infantless group was nonchalant (which does describe mangabeys). During the following month an S.B. female gave birth to an infant. During 4.5 days (54 hours of vigil) in mid-May of 1977, S.B. group spent ~27 hours foraging in *F. dawei* #4. Chimpanzees spent 11 hours during that vigil period in the same tree. Upon the apes' arrivals the new mother always moved out of the tree. S.B. group's previous nonchalance was replaced by a contrasting wariness. The female and her new infant had become the social focus of the group and the nucleus of grooming clusters. She repeatedly left the *dawei* when the apes arrived. The group's response to chimpanzees had changed. Although some unknown event may have changed the group's relationship with chimpanzees, the behavior of the new mother and the group suggest that the infant was the catalyst. Although anecdotal, this further suggests that mangabeys consider chimpanzees to be dangerous rather than neutral. In general, adult chimpanzees always seemed capable of dominating mangabeys, while adult male mangabeys dominated juvenile chimpanzees.

Red colobus. Chimpanzee interactions with red colobus appeared unidirectional at Ngogo. During all 15 passive or actively antagonistic confrontations red colobus invariably displaced or thwarted the approach of chimpanzees. Four times dominance was passive in the sense that the approaching group of monkeys did not appear intent on displacing chimpanzees (and may have been unaware of their presence); nevertheless the apes glanced at the monkeys repeatedly, then abandoned food trees before I thought they would have done otherwise.

On 11 other occasions (1 instance/44.3 hours of observation on chimpanzees) red colobus threatened, chased, and/or attacked chimpanzees causing the latter to retreat or flee from a food resource. Because such interactions are rare in the literature on chimpanzees, each incident from my field notes is included below.

24 March 1977, 0945, 10 m N.W. *C/8.5*. A solitary young adult female chimpanzee looks behind her, W.S.W., toward approaching red colobus as she ascends W. route of *F. m.* #9. She makes a quick circuit to N.E. route eating only a few figs along the way, and "settles" in lowest portion of *F. m.* crown above N.E. route, apparently not feeding. Meanwhile, an adult male red colobus has moved up W. route, followed the chimp's route, but stopped halfway [through the crown], about 8 m from her, and "chists" [vocally threatens] rapidly at her while advancing toward her in short, abrupt pairs of steps. After 1 min the monkey rushes 4–5 m toward her, and she descends 2–3 m. The red colobus continues threatening. After 30 sec more, he rushes toward her again, chisting, and she descends 3–4 m. *As* she descends, the red colobus "hurls" himself through the air down N.E. route after her, chisting with a great crashing of foliage. She screams 2–3 high-pitched, short shriek-screams and rapidly descends to the ground and moves away. The red colobus climbs back up into *F. m.* #9. [During this period a few redtails were in the tree with a solitary adult mangabey: none were involved in the interaction.]

6 May 1977. 1828–1842, 30 m W 850 *F*. [In response to my presence a JF (4–5) chimpanzee, Fern, is hiding 25 m up in a *Lovoa swynnertoni,* and her presumed older sibling, Fearless, a SAM [Subadult Male], is 35–40 m S. of her. Fearless intermittantly screams and pant-hoots, apparently at me, until 1839. By this time adult males from a nearby group of red colobus have nearly surrounded Fern and are chisting at her while ignoring me.] 1841–42. Branch stomping and barking screams lasting ~30 sec by Fearless, who has climbed back toward Fern. Absolute silence follows his display. 1843. Fearles and Fern [Whom I do not see well] move S., away from the red colobus. Note: Fearless returned to Fern as if to "rescue" her from the red colobus.

14 November 1977, 1647, center 1740 *F*. A chisting adult male red colobus enters *F. m.* about 6 m from adult female chimpanzee, Ita, who moves 5 m away. Within 5 sec, the red colobus rushes toward Ita, who screams repeatedly while rushing toward E. route. The red colobus chases her and is joined by two other adult males from his group, which is moving in from N.E. Ita continues to flee, pausing only long enough to glance over her shoulder at the aggressive red colobus. She makes reckless, headlong leaps down into the foliage from tree to tree using her only hand to grab hold [Ita is missing her left hand], Zira [adult female] rushes over *toward* the chase scene, screaming hoarsely, but Ita is almost on the ground before she arrives. Both females disappear at 1648 while 3–4 red colobus males chist above them. Owl [third female ape present] was not seen to depart, but now is gone.

6 February 1978, 1057, 3 m N 1840 9. Owl arrives at base of F. m., climbs about 5 m up, sees me, hoos, then sits and watches me [~16 m away]. . . . Several red colobus are in F. m., 1058. 2 adult males move 15 m W. to perch in a Chrysophyllum albidum above Owl and gaze down at her. They do not feed. One male notices me, moves 5 m away and chists. 1113 [Owl has been waiting beneath tree quietly] both males move 3–4 m closer, chist and branch shake at Owl [and me?]. [Owl continues to sit until 1209 then climbs down and walks N.]

16 March 1978, 1123, 25 m NNW C.5/8.5. [Solitary adult male, Stump, climbs into a Pseudospondias microcarpa heavily laden with fruit. 40 m S. red colobus males are already chisting and branch-shaking toward Stump, who forages despite them. For 12 min the red colobus move closer and intensify their threats. 1135, red colobus males are now within leaping distance of the semi-isolated Pseudospondias, 10 m from Stump, who climbs one-handedly to the ground and moves off.]

17 March 1978, 1040, near junction 8.5/F. I hear a single chimp scream, sounding like a juvenile. . . . At 1045 I hear another low-intensity scream, juvenile type. 25 m through the understory a large adult male chimp walks calmly along a fallen log. A juvenile male chimp about 8 years old, sits on the log near where the male had just been. He looks about as if to check if anyone is watching him. 1048, he follows the adult male. Within 2 min red colobus begin alarm calls about 50 m E.N.E. of the log. At 1054 I hear the sounds of animals rushing toward me from E. Two chimps appear. The smaller one disappears on a W.N.W. course. Behind him I see an adult male red colobus climb up from the ground into a sapling. More chists at 1058. The large chimp rushes back, E.N.E. along his original path. About 10 m behind him runs an adult male red colobus and behind him runs a second red colobus; both pursuers on the ground! All three disappear E.N.E. into the foliage. No chimp vocalizations since 1045. None of the primates seemed to notice me. At 1113 a sonorous buttress drumming from about 100 m N.N.E. Red colobus chist again.

30 April 1978, 1138, 7 m E. D.5. ([Fearless, a SAM, and Owl both have been feeding on figs of F. mucuso #2 for 53 and 31 min respectively.] One or two red colobus adult males move into F. m. # 2 from S., chisting as they move closer. Fearless and Owl simultaneously move S. then exit to S. E. with 1–2 red colobus rushing toward them. Both chimps descend to the ground; there is a brief scuffle in the exit tree about 8 m up, with much colobus vocalizing, then quiet.

1 May 1978, 0841, 70 m N.W. 9/E.5. [Owl has been feeding on figs of F. exasperata for 14 min.] I hear red colobus chisting to the N.N.W. At 0843 several of them are moving rapidly S.S.E. toward Owl. She continues to forage until 0848, at which time 3–4 adult male red colobus have reached the N.W. tree nearest to the F. exasperata and are chisting and looking at Owl. She descends to the ground 15 m from me, but is unaware of me, then walks N. Within 15 min the red colobus have moved through the F. exasperata and beyond.

9 May 1978, 1108, junction 8/K. [At 1046 I arrive at F. mucuso #6, during

census walk and find Stump, Fearless, an old adult female, a young adult female, and a juvenile (the last 3 unidentified) eating figs. Within 50 m are groups of mangabeys, blue monkeys (both of which have members in *F. m. #6*) redtails, and red colobus.] 1108–1119, red colobus file into *F. m. #6* from N.W. All 5 chimps move to S. side of crown. [The young adult female rushes from the tree at 1117 and gives short screams.] An adult male red colobus rushes toward the old female chimp. They grapple about 3 sec, then the female breaks it off by leaping 3–4 m down into an understory tree. By 1120 all 5 chimps are out of, but very close to *F. m. #6*. They pause and give sharp screams, Stump giving a "wraagh-bark." The red colobus chist. The chimps move slowly into *F.m. #5*, 40 m away and without fruit. Stump wraagh-barks 2 more times in next 2 min, accompanied by chimp screams that elicit chists and branch shaking by the red colobus adult males.

29 May 1981, 0833, 5 m E 710 *E*. A solo chimp climbs into N.E. route to enter *F. m. #1*. ≥ 3 red colobus adult males rush toward it from *F. m. #1* and vocalize. The chimp moves away. Twenty sec later the colobus advance again and the chimp [whom I still have not seen clearly] hastily descends to the ground 15 m from us to land with a thump. It disappears. The colobus vocalize another 5 min.

31 May 1981, 1025, center 1265 *5.5*. [As we move closer toward juvenile male, Ashly, he screams several times, eliciting mangabey "chitters," blue monkey "pyows," redtail "hacks" and "chirps," plus red colobus threat vocalizations. ≥ 3 red colobus adult males rush toward Ashly from N. He moves 15 m S. and continues screams. The colobus move to the edge of Ashly's tree but do not enter it.]

On another occasion (28 May 1981, 1000) when no red colobus were present, one of the most dominant adult male chimpanzees at Ngogo, Eskimo, carried a large ventral section of skin, which, judging from its size and color, came from a freshly killed red colobus. Eskimo nibbled bits of flesh from the skin and alternated them with bites of figs from *F. mucuso*. About 48 hours later (30 May 1981, 0930), Eskimo was still (?) carrying a somewhat smaller skin of the same appearance, but he no longer nibbled at it. It is not known whether Eskimo killed, scavenged, or robbed from another predator the prey item (see Morris and Goodall 1977 for a discussion of prey thefts by chimpanzees).

In Gombe chimpanzees were reported to attack, kill and eat red colobus (Goodall 1968; Wrangham 1975; Busse 1977; Riss and Busse 1977). Busse (1977) reported that chimpanzees killed 39 red colobus between 1973–74 at a rate of 0.95 kills/100 hours of observation.

During thousands of hours of observation of red colobus in Kibale Forest, neither T. Struhsaker nor L. Leland have seen one killed by a chimpanzee. Although, as T. Struhsaker (pers. comm.) suggested, time spent with a prey species is less likely to result in observations of predation than time spent with predators. Both T. Struhsaker and L. Leland have seen probable attempts which failed, possibly due to their presence. Busse (1977) suggested that this lack of predatory incidents at Kibale was due to the chimpanzees not being habituated to human observers. This argument has merit but does not explain why I only once saw evidence of predation during nearly 488 hours of observation of the predators. During several of the chimpanzee—red colobus interactions I observed, none of the primates were aware of my presence.

Contrary to an impression that may be gained easily, Gombe chimpanzees were not always dominant over red colobus. After conducting a study of red colobus there and seeing only one attempt at predation, a failure, Clutton-Brock (1972:58) noted, "On several other occasions, the red colobus harassed chimpanzees." Harassment included a male chimpanzee being chased by two adult male red colobus, one of whom was on the ground.

Busse (1977) reported that 75.3 percent of all chimpanzee encounters with red colobus at Gombe during 2 years led to attempts at predation; nearly half the attempts were successful. Males were the primary predators (McGrew 1979). In contrast, none of the 15 encounters I saw during nearly two years in Kibale led to successful predation, but only one of these encounters (17 March 1978) included an adult male chimpanzee unhandicapped by wounds. The aggression directed at chimpanzees by red colobus in Kibale appeared to be antipredator behavior. What factors account for the difference in rates of predation between Gombe and Kibale?

Despite their reputation for abandon, chimpanzees were cautious when more than 15 m above the earth. The benefit of attacking monkeys, who are more agile and immune to the effects of falls, while in the canopy of the forest may not be worth the risk. Busse (1977) also reported that during 12 of 33 unsuccessful attempts at predation in Gombe red colobus were seen chasing chimpanzees through the canopy. Wrangham (1975) suggested that red colobus at Gombe were most vulnerable to chimpanzees when caught in low, broken canopy. Chimpanzees also killed redtails (Kawabe 1966; Wrangham 1975), but at Gombe redtails spent less time in broken canopy than did red colobus and were preyed upon less frequently. Areas of low growth and broken canopy seemed to be less frequently used by Kibale chimpanzees than by their Gombe counterparts, but it was in such a habitat in compartment 30 where L. Leland (pers. comm.) twice saw a chimpanzee ascending small isolated trees containing fleeing red colobus during an unsuccessful attempt at predation.

Busse (1977) proposed that the multimale, large group structure of red colobus was an antipredator adaptation, noting that they usually were sympatric with chimpanzees. The evolutionary and ecological determinants of group size and structure among primates have been discussed by Crook and Gartlan (1966); Struhsaker (1969); Crook (1970); Kummer (1971); Eisenberg et al. (1972); Goss-Custard (1972); and Clutton-Brock and Harvey (1977). These workers emphasized that a constellation of selective pressures have shaped present primate social systems (see chapter 7). A comparison of rates of predation by chimpanzees on large and small group of red colobus, containing varying numbers of adult males, is necessary to test Busse's (1977) hypothesis. Data from Gombe and Kibale do show incontrovertibly that adult male red colobus are able to drive away chimpanzees in various habitats. They also attack humans. J. Skorupa (pers. comm.) was attacked by several males in secondary growth north of Kanyawara. He lost his footing and had to drive several off with a stick. Another adult male monkey leapt upon the back of a Kanyawaran, Kiisa, (pers. comm.) and bit him as he attempted to drive red colobus from some trees. It seems likely that a one-male group would be more vulnerable to predation than a group with several males, but other factors must affect vulnerability; redtail, blue, l'Hoest's and black and white colobus monkeys all live in one-male groups and apparently are preyed upon less frequently than red colobus, though they are less numerous as potential prey. Further research will, I hope, clarify the nuances of the evolutionary forces affecting interactions between red colobus and chimpanzees.

Black and white colobus. I rarely saw black and white colobus in close proximity to chimpanzees (see column e, table 22, below). The single interaction I saw consisted of an adult male colobus briefly charging a solitary juvenile male chimpanzee, who rushed several meters downward in their shared tree, then paused for a moment before dropping to the ground and walking away.

Red duiker. I saw interactions between chimpanzees and antelope only indirectly. On three occasions a red duiker rushed toward me, oblivious of my presence (one stopped from a dead run only 0.5 m short of me), coming from a direction from which chimpanzees arrived only seconds later. By comparison with the behavior of undisturbed duikers, these were obviously fleeing.

I observed an adult and a subadult male baboon chase a female red duiker with her newborn kid, but because the individuals were not habituated to me, I did not see the outcome of the chase. On another occasion I found the twisted skin of what appeared to be a blue duiker in an area recently frequented by chimpanzees and baboons. L. Scott (pers. comm.) told me that the skin resembled exactly the remains of antelope kills made by baboons in Gilgil (see Strum 1975). This circumstantial evidence suggests that duikers may be preyed upon by both species of primates.

Elephant. I observed chimpanzees in close proximity to elephants only once. During a session of mutual grooming between two adult females, Gray and Zira, a group of four to ten elephants rushed past me to either side. As soon as the pachyderms had passed I looked toward Gray and Zira, who still were in the same location and quietly grooming as before. A second rush of 19 to 26 elephants occurred 2.58 hours later, but the apes had departed 10–18 min before. They returned 35 min later; whether they had avoided the second group of elephants purposely or inadvertently is unknown.

Frugivorous birds. Chimpanzees rarely tolerated large frugivorous birds within 5 m of them while feeding. Often a feint or a lunge toward a foraging bird sent it to a more distant section of the crown. Black and White-casqued Hornbills were the least tolerated species. Great Blue Turacos also were chased frequently. Even medium-sized birds such as the Green Pigeon (*Treron australis*) sometimes were chased.

On the other hand, chimpanzees, especially juveniles, sometimes ducked their heads or flinched when a gliding Great Blue Turaco swooped close over their heads before alighting in a fruit tree.

Humans. At Ngogo chimpanzees fled from me during my early months of research. The first individuals to exhibit tolerance for me were adult females with large perineal swellings and without accompanying offspring. As the project progressed many individuals gained a tolerance for me, although only two (of eight) adult females with infants at Ngogo were among them. Chimpanzees who tolerated me seemed to be much less tolerant of the Ugandan trail cutters in the study area, although they showed a relatively high degree of tolerance for other European observers (e.g., K. G. Van Orsdol) to whom I introduced them, if those observers practiced protocols signifying nonaggressive intent.

Some unhabituated chimpanzees, especially mothers with infants, apparently were too afraid even to climb down from a tree to flee. Sometimes they displayed vocally and rushed through the crown, kicking off dead limbs, sending them crashing down.

During the ninth month of my study, while observing an old female, Gray, and two young females, Zira and Owl, feeding together in a *F. natalensis*, lightning struck very near the crown of their tree. Thunder was virtually simultaneous. Gray, who had been extremely tolerant and usually ignored me, stared at me intently, scrutinizingly, as if to determine exactly what I was doing. After 2 sec of scrutiny, Gray rushed out of the fig tree to ''hide'' 80 m away. Neither remaining young female exhibited a startled response to the close lightning, and both continued to feed as if *nothing* had happened. Yet all three chimpanzees had been feeding within 5–8 m of one another at the time. Seven months passed (six contacts) before Gray tolerated me again as she had done before the lightning episode.

Until 14 years before this study Bakonjo poachers had apparently employed firearms in their poaching of primates in Kibale Forest (S. M. Yongili pers. comm.; hearsay on his part). Gray's sudden transformation to suspicion and fear during the thunderstorm suggests that she had a mental association of humans, thunder-like sounds, and danger. From her wrinkles, grizzled hair, and postmenopausal state, I estimated her age to be about 35 or 40 years. I suspect she had some memory of humans hunting with firearms and also may have remembered some of the consequences of their hunting. As mentioned in chapter 3, a local informant reported that Batoro poachers occasionally were trapping and killing chimpanzees north of compartment 30 during this study.

Competition

Competition between individuals of one or more species is a part of nature which seems intuitively obvious (Darwin 1859), yet complete documentation of the competitive relationship between sympatric mammals is normally an awesome challenge. The following treatment is merely an outline suggesting the nature of interspecific competition from the perspective of chimpanzees at Ngogo.

Table 22 summarizes the ecological parameters (percent of dietary overlap of food types eaten by each species of monkey [Note that these figures are partial at best] and each one's biomass density) I used to compute a synthetic index of competition (column c). This index predicts a rank order for frequency of displacements between species. The observed rank order is included in column d. The index in column c indicates that redtails, red colobus, and baboons were competitors to be reckoned with almost equally from the perspective of a chimpanzee. Mangabeys were the only other serious competitors among the primates. These four species (including *ascanius-mitis* hybrids) accounted for 97 percent of displacements involving chimpanzees.

Originally I assumed that an increased amount of time spent in interspecific association would lead to an increased frequency of dominance interactions between species. But this assumption did not account for interactions between species which are virtually incompatible, such as red colobus and chimpanzees, the interspecific combination which led to the most frequent dominance interactions. Although increased time together did lead to increased dominance interactions, it also reflected percent of dietary overlap. Types of interactions between species which spend time together vary depending upon whether they are in association because their foraging niches are mutually exclusive to some degree but their association confers other benefits (such as between redtails and red colobus [Struhsaker 1980]), or because they are attracted to the *same* food resource and are in competition (see table 18). Because of these considerations, I decided against using time spent together as a predictor of interspecific interactions.

This analysis of interspecific competition would have been improved had data been available to compute dietary overlap for each sympatric species by using a summation of the proportions of the quantities of each food type it consumed in its total diet to those quantities consumed in the diet of the apes. But comparison of the number of shared food types, even while ignoring their proportions in total diet, did indicate competition, though not its intensity. The percentage of dietary overlap used here combined with biomass density yielded a competitive index and dominance predictor which agreed with the interspecific dominance interactions I observed.

TABLE 22. PARAMETERS OF COMPETITION BETWEEN CHIMPANZEES AND SYMPATRIC ANTHROPOID PRIMATES AT NGOGO, KIBALE NATURE RESERVE, UGANDA.

Species	(a) % dietary overlap with food types of chimpanzees	(b) Biomass density (kg/km²)	(c) Predicted rank of displacement, synthetic index = (a) × (b)	(d) Observed rank of displacement (apes dominant)	(e) % obs. time spent in assoc. with chimpanzees
Redtail monkey	48	255.1	122.4	2 (11)[a]	5.81
Red colobus	20	545.8	109.16	1 (0)	0.13
Baboon	24	432.3	103.8	4 (2)	0.52
Grey-cheeked mangabey	38	127.0	48.3	3 (≥5)	6.54
Blue monkey	36.5	4.7	1.67	6.5 (0)	0.89
Black and white colobus	2	28.0	0.56	5 (0)	0.003
l'Hoest's monkey	8	1.8	0.14	6.5 (0)	0.05

NOTE: Data are used to predict a rank of displacement between chimpanzees and monkeys.
[a] Observations of chimpanzees and redtails and chimpanzees and ascanius-mitis hybrids were pooled because of the social affinity of the ceropithecine spp.

Despite dominance interactions, the primary mode of competition seemed to be exploitative rather than interference (Elton and Miller 1954; Park 1954; Miller 1967). In the arena for scramble competition (Nicholson 1957) which is Kibale Nature Reserve, the first individuals to find a food resource exploited it. This simple picture was complicated by different tolerences between species for unripe fruit. Chimpanzees preferred apparently riper fruit and seemed to visit fruit trees to ascertain the ripeness of their crops (see also Wrangham 1975). Red colobus repeatedly spent hours in *F. mucuso* trees harvesting a large amount of unripe fruit, consuming only a fraction, and dropping the rest in a state that chimpanzees found unpalatable. In this way the colobus had a distinct competitive edge over the apes. Interference competition did occur as one species supplanted another. Chimpanzees were dominant over the four species whose overlap in food types exceeded 20 percent, but, as Emlen (1973:307) noted, "active exclusion will not always occur, for it may be that the advantage of ousting a competitor is more than offset by the cost of ousting him." In addition, the interplay of feeding competition and antipredator strategy confounds the use of dominance interactions as a strict indicator of competition.

The ability to displace competitors can have an important influence on foraging strategy. In their analysis of foraging strategies in a patchy environment, MacArthur and Pianka (1966:607) concluded, "when the gain to a jack-of-all-trades in reduced travelling time makes up for his lower efficiency compared to the patch specialists, then the jack-of-all-trades will out compete both specialists." Chimpanzees are patch specialists to a greater degree than are *Colobus* and *Cercopithecus* spp. in Kibale, and probably are more so than mangabeys (see Waser and Floody 1974), and the apes range over a much wider area than do single groups of any of the species of monkeys studied in Kibale. Interspecific competition at some patches was so severe that chasing and fighting between species determined who remained to forage in a patch. Being a consistent winner in such encounters can help tip the balance of advantage in foraging from the jacks-of-all-trades, *Cercopithecus* spp. and mangabeys, relatively speaking (Struhsaker 1978), to the specialist. Chimpanzees are large-bodied specialists in a patchy environment, and they compete with small-bodied, sympatric primates that outnumber them 197:1. If chimpanzees could not displace monkeys, their niche might be untenable.

6

SOCIAL STRUCTURE

Perhaps the most intriguing and recondite aspect of chimpanzee natural history is social structure. A chimpanzee may be solitary on one day, may travel with two others the following day, then be with a third individual on the next. Associations between individuals tend to be protean, complicated, and confusing. At first view the coherent and permanent group structure so characteristic of other anthropoid primates appears to be totally lacking. Superficially chimpanzees appear to be footloose; they gather during times of plenty but spent most of their time in small parties and in solitude. That behind this apparent randomness is a tight social order can be only imagined.

Analysis of social interactions belies this superficial disorder. Sociality among chimpanzees is far from random. Instead it is a highly evolved system that allows the individual the best of both worlds: rich sociality combined with low competition between individuals for scarce resources. This chapter examines social behaviors and their probable functions and leads to a theoretical discussion of the social ecology of chimpanzees in the chapter following.

One inadvertent, but fortuitous result of the program of provisioning bananas to the apes of Gombe National Park was the experimental demonstration that chimpanzees travel in larger than normal parties when normal constraints of foraging are lifted (Wrangham 1975). At Ngogo affiliative behaviors, such as mutual grooming, were far more frequent than agonistic interactions (see below). Both of these observations indicate that companionship among chimpanzees may be favored over a solitary condition. This is not a trivial inference; it suggests that sociality is constrained by feeding ecology. Food resources and feeding ecology seem to be paramount aspects in a chimpanzee's day-to-day life. If feeding and minimizing intraspecific competition for food were the only issues, most chimpanzees probably would be solitary except when the need to mate or care for offspring dictated some socialization. If sociality is "expensive," we would expect individuals to direct their social activities so as to derive the maximum potential individual and/or inclusive fitness from them. Traveling companions are good indicators of a chimpanzee's choice of apportionment of social interactions because it must pay for companionship by increased competition during feeding.

ASSOCIATIONAL TENDENCIES

Traveling parties

Choice of traveling companions represents a decision based upon the benefits of being with certain other individuals versus the cost of competing with them. Because chimpanzees often are solitary (see below), one may conclude that the nonmaternal-dependent traveling companions of an adult are companions due in large part to the *choice* of that adult, rather than to necessity or to the mores of a tightly knit social group that dictate close proximity. Given this choice, it seems clear that a chimpanzee travels with another for its *own* benefit. Thus the size and composition of traveling parties are phenomena determined by individual behavior for the benefit of the individual. This perspective is implicit in the following analyses.

A traveling party is defined here as one or more chimpanzees who arrive at or depart from a particular place, traveling in the same direction, within 3 min of one another. Wrangham (1975) used a same-path-within-15-min criterion for defining traveling parties of Gombe chimpanzees, but this was in a special environment, the provisioning area. The 3-min criterion used here I judged would be less likely to include individuals traveling separately as one party. Thus it is more conservative.

Size. The average size of all traveling parties ($N = 667$) of chimpanzees I counted in Kibale Forest was 2.6 individuals ($R = 1$–24). Figure 15 summarizes frequencies of party sizes of mixed and unisexual composition I saw during the first 17 months of this study. The modal size was 1 for both sexes. Solitary males accounted for 12.9 percent and solitary females accounted for 20 percent of my observation *time* on each sex between 1976 and 1978. If I considered a mother and her infant as a solitary unit, but older accompanying offspring as companions, 44.3 percent of all the parties I saw in Kibale were solitary. The largest parties were of mixed sexual composition, as would be expected by chance ($\bar{x} = 5.4$, $N = 104$). The average size of parties containing males only was 1.7 ($N = 178$), while those containing one or more adult females only, with or without attendant offspring, was 2.5 individuals ($N = 317$). The major difference between parties containing either male or female adults

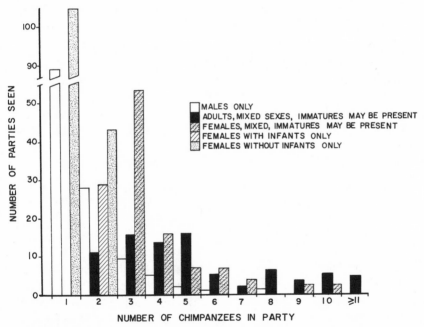

FIGURE 15. Frequencies of sizes and sexual compositions of traveling parties of chimpanzees seen at Ngogo between December 1976–May 1978.

was the presence of infants and young juveniles in the latter, which escalated the head counts of the parties but often added little biomass.

In general, sizes of traveling parties of chimpanzees were small. I had the impression that this was related to the availability of superabundant resources. A month-by-month comparison of average party size versus availability of fruit from selected large trees (figure 14) suggests a trend, but one that is not statistically significant. More complete data on food availability might strengthen this.

Table 23 presents a different perspective on sizes of traveling parties. I tabulated all episodes of travel that I witnessed among 5 of the 7 Ngogo adults whom I contacted most frequently. Among these adults solitary travel accounted for 46 percent of all travel episodes ($N = 267$). The two very old adults, Gray and Stump, showed a dramatic increase in their tendency to travel with others rather than solitarily compared to the younger adults ($\chi^2 = 28.788$, d.f. $= 4$, $p \leqslant 0.0005$). Reasons for this difference are not clear. However, old individuals seem more likely than young ones to be included in and to engage in long-standing interindividual associations simply because of the additional time they have had to form such bonds. Another consideration is that Gray was postreproductive. Perhaps she could afford the increased competition for food incurred by traveling with others because her future reproductive interests were small. Perhaps Gray's only chance of increasing her reproductive success was to travel with individuals who may have been her offspring and relate to them in ways that improved their ability to forage, etc.

Wrangham (1975) found that male chimpanzees at Gombe traveled solitarily in 8.3–41.8 percent of all parties during 2 dry seasons and in 36.2 percent of all parties during one wet season. He found also that party sizes varied between habitats and somewhat between seasons. During a 50-day follow of the alpha-male chimpanzee, Figan, of Gombe's Kasakela community, Riss and Busse (1977) found that he spent only 16 percent of his time alone (a low figure for Gombe) and that mean group size was 4 individuals, excluding subadolescents. The largest parties were observed at abundant food resources. Izawa and Itani (1966) reported 20 percent of chimpanzee parties in Kasakati Basin consisted of solitary individuals, while average party size was 6.8 individuals. Jones and Sabater Pi (1971) reported that they never were certain of having seen a solitary chimpanzee in Rio Muni, but group sizes averaged 4.66 ($N = 3$) in

TABLE 23. SUMMARY OF SIZES OF TRAVELING PARTIES CONTAINING 5 OF THE 7 MOST FREQUENTLY SEEN, INDIVIDUALLY RECOGNIZABLE CHIMPANZEES AT NGOGO, KIBALE FOREST, UGANDA.

Age-sex reproductive class of individual	No. of days seen	No. of sightings	No. of hours observed hour:min	No. of times seen traveling	Size of traveling party						
					Solitary	With 1 other	With 2 others	With 3 others	With 4 others	With ≥5 others	$\overline{X}$ size
Adult female, prime, Blondie, with infant, Butch, and juvenile, Bess.	20	28	26:32	46	26[a] 56.5%	5 10.9%	3 6.5%	3 6.5%	4 8.7%	5 10.9%	4.37 2.37[b]
Adult female, old, Gray	32	35	67:04	57	17 29.8%	25 43.9%	8 14.0%	3 5.3%	2 3.5%	2 3.5%	2.28
Adult female, young, Owl	32	39	82:48	51	37 72.5%	3 5.9%	4 7.8%	3 5.9%	1 2.0%	3 5.9%	1.92
Adult male, young, Raw Patch	40	60	69:18	81	34 42.0%	17 21.0%	9 11.1%	10 12.3%	7 8.6%	4 4.9%	2.49
Adult male, old, Stump	20	23	35:12	32	8 25.0%	10 31.3%	6 18.8%	5 15.6%	1 3.1%	2 6.3%	2.56

a "Solitary" considers Blondie, Butch, and Bess as a single mother-offspring unit.
b Mean size of traveling parties considering Blondie, Butch, and Bess as a single mother-offspring unit.

one area and 11.2 (N = 5) in another. Nishida (1968:183) found that 30.7 percent of chimpanzee parties in Mahale Mountains were solitary, considering mothers with accompanying offspring as units (N = 218, R = 1–28), noting in additon that sizes of parties varied seasonally. Reynolds (1963) and Reynolds and Reynolds (1965) did not report solitary chimpanzees or mean sizes of parties in Budongo but they did report a traveling party they conservatively estimated at 62 individuals! They also noted that party sizes changed seasonally. Sugiyama (1968) reported that 28 percent of traveling parties at Budongo (N = 514) were solitary chimpanzees ($\bar{x}$ = 3.9 apes/party) and that mean size of parties changed seasonally.

Clearly sizes of traveling parties of chimpanzees vary greatly within a single habitat, between seasons, between habitats, and between observers. How much of this variation is due to differing techniques of investigation and/or analysis is difficult to determine. Solitary apes, for instance, are difficult to spot. More reports of solitary individuals might be expected from observers in long-term studies because they probably have developed superior bushcraft. Chimpanzees in provisioned areas (Mahale and Gombe) seemed more likely to travel with companions.

Composition of parties. Table 24 summarizes the associational tendencies of age-sex classes of chimpanzees in Kibale Forest. Each individual of a specific age-sex class, in the company of another, counted as one incident. Two adult males traveling together received a score of 2 because each of them was traveling with another adult male, and so on.

Both adult and juvenile males were more likely to travel with other males than with females. Adult males preferred one another and differed significantly from adult females unaccompanied by infants in the sex of their traveling companions (χ^2 = 56.793, d.f. = 7, $p \leq 0.005$), and differed in the same way from females accompanied by infants (χ^2 = 96.196, d.f. = 7, $p \leq 0.0005$). Juvenile males also differed from juvenile females in sex of traveling companions, each preferring like sexes (χ^2 = 23.062, d.f. = 7, $p \leq 0.005$), which indicates that sexual preferences begin early in males' and females' postmaternal-dependent ontogeny.

Surprisingly, even adult females with infants differed significantly from adult females without infants in their traveling companions. Females with infants were more likely to travel with other females with infants and with juvenile females than were females without infants, while the latter

TABLE 24. SUMMARY OF AGE-SEX CLASS ASSOCIATIONAL TENDENCIES OF TRAVELING PARTIES OF CHIMPANZEES AT NGOGO, KIBALE FOREST, UGANDA.

Age-sex class	No. of times seen traveling	Seen in party with one or more of the age-sex classes below						
		Adult male	Subadult male	Juvenile male	Adult female without infant	Adult female with infant	Juvenile female	Juvenile, undetermined sex
Adult male	250	134 (53.6)	25 (10.0)	58 (23.2)	77 (30.8)	38 (15.2)	22 (8.8)	12 (4.8)
Subadult male	31	16 (51.6)	—	13 (41.9)	18 (58.1)	9 (29.0)	6 (19.4)	3 (9.7)
Juvenile male	104	46 (44.2)	17 (16.3)	16 (15.4)	30 (28.9)	38 (36.5)	13 (12.5)	6 (5.8)
Adult female without infant	298	75 (25.2)	24 (8.1)	41 (13.8)	139 (46.6)	57 (19.1)	63 (21.1)	24 (8.1)
Adult female with infant	167	33 (19.8)	10 (6.0)	38 (22.8)	54 (32.3)	44 (26.3)	69 (41.3)	25 (15.0)
Juvenile female	105	18 (17.1)	8 (7.6)	12 (11.4)	45 (42.9)	56 (53.3)	18 (17.1)	2 (1.9)

NOTE: Lower (quantity) represents percent of total times seen.

were more likely to travel with adult males and/or other adult females without infants than were those with infants ($\chi^2 = 24.519$, d.f. $= 7$, $p \leq 0.005$). All classes of females were significantly more likely to travel with other females than with males.

The tendency to travel in parties of nonrandom age–sex composition was tested further using subsets of specific data. The first subset tests a situation in which larger than normal party sizes were usual. Table 25 summarizes the associational tendencies of 28 individually recognizable chimpanzees beyond infancy who visited *F. dawei* #4 in April 1978 (see also figure 10). Again, males traveled more with other males than with females, while females overwhelmingly preferred other females as traveling companions ($\chi^2 = 60.26$, d.f. $= 1$, $p \leq 0.0005$).

The second subset (table 26) is a summary of the sexual representation of all traveling companions beyond infancy of five chimpanzees whom I contacted frequently at Ngogo during the first 17 months of this study. Raw Patch and Stump were significantly more likely to travel with one or more other adult males than were the three adult females, who were more likely to be with one or more other females ($\chi^2 = 24.776$, d.f. $= 1$, $p \leq 0.0005$). An analysis of all known traveling companions, not considering numbers of times seen traveling, indicates again that Raw Patch and Stump were more likely to travel with another male than a female, and that Blondie, Gray, and Owl were more likely to travel with another female than with a male ($\chi^2 = 42.112$, d.f. $= 1$, $p \leq 0.0005$). Another interesting comparison in table 25 is the representation in "num-

TABLE 25. SUMMARY OF SEXUAL CHOICE OF TRAVELING COMPANIONS AMONG POST-INFANT-AGED CHIMPANZEES VISITING AND DEPARTING A FRUITING *FICUS DAWEI* (SPECIMEN #4) BETWEEN 16–26 APRIL 1978, AT NGOGO, KIBALE FOREST, UGANDA.

Sex of visitor	Individual traveling companions beyond infancy to and/or from tree		
	Male	Female	Total sexes
Male, N = 12; 53 visits	103	79	182
Female, N = 16; 86 visits	72	250	322
Total; N = 28; 139 visits	175	329	504

TABLE 26. SUMMARY OF SEXUAL REPRESENTATIONS IN PARTIES OF TRAVELING CHIMPANZEES CONTAINING 5 OF THE 7 MOST FREQUENTLY SEEN INDIVIDUALS AT NGOGO, KIBALE FOREST, UGANDA.

Age-sex reproductive class of individual	No. of times seen traveling	No. of times seen traveling with ≥1 individual beyond infancy		Total no. of traveling companions beyond infancy of known sex			No. of infants traveled with
		Male	Female	Total	Male	Female	
Adult female, prime, *Blondie*, with infant, *Butch*, and juvenile, *Bess* [a]	46	8 17.4%	18 39.1%	44	9 20.5%	35 79.5%	14 24.1%
Adult female, old, *Gray*	57	9 15.8%	36 63.2%	68	13 19.1%	55 80.9%	5 6.8%
Adult female, young, *Owl*	51	6 11.8%	13 25.5%	42	9 21.4%	33 78.5%	6 12.5%
Adult male, young, *Raw Patch*	81	33 40.7%	23 28.4%	97	51 52.6%	46 47.4%	13 11.9%
Adult male, old, *Stump*	32	21 42.9%	9 28.1%	47	31 66.0%	16 34.0%	1 2.0%

[a] *Blondie*, *Butch*, and *Bess* are considered here as a single mother–offspring unit. Note that the last row entry, 14 and 24.1% does not include *Butch* as an infant traveled with.

ber of individuals traveled with'' made by dependent infants not belonging to the focal chimpanzee. I saw Stump traveling in the company of an infant (and its mother) only once during 24 nonsolitary episodes of travel, while Blondie, who carried her own infant, traveled with 14 other infants during 20 episodes of nonsolitary travel. These data are in concert with the tendencies for mothers with infants to prefer other mothers with infants as companions over all other age-sex-reproductive classes.

Azuma and Toyoshima (1962) reported the ''familoid'' group, containing one adult male plus one or more females with accompanying offspring, as the most common social grouping of chimpanzees at Kaboko Point, Tanzania. Goodall (1965) reported that 30 percent of groups ($N =$ 350) were of mixed sexes, 28 percent were of males only, and 24 percent were of females with accompanying offspring. But because ''virtually all chimpanzee observations [at Gombe between 1961 and 1969] were at the provisioning station'' (Busse 1977:909), Goodall's (1965) mixed groups may have been overrepresented due to the presence of large amounts of food. Halperin (1979) found that adult males and adult females with infants at Gombe were more likely to be in unisexual parties or solitary than in mixed parties. Kortlandt (1962) distinguished between two major types of traveling parties; ''sexual groups'' containing adults of both sexes, and ''nursery groups'' composed of females and immatures. Nishida (1968) recorded of 218 subgroups that 10.6 percent were all-male ($\bar{x} =$ 2.6 apes/party), 51.8 percent were mixed ($\bar{x} = 13.1$ apes/party), 13.3 percent were of mothers and offspring ($\bar{x} = 5.2$ apes/party). But here again provisioning may have biased these data toward large parties of mixed composition. Reynolds and Reynolds (1965) reported of 103 bands that 13.6 percent were of males only, 70.9 percent were mixed, and 15.5 percent were of mothers. These latter data are biased toward large parties because the Budongo chimpanzees usually were located by their vocalizations; and larger parties (especially containing males) vocalize more than smaller ones (see also section on food-calling, below).

Kortlandt (1962), Sugiyama (1968), Nishida (1979), Pusey (1979) and the present study contribute to the recurrent theme of males exhibiting more affiliative behavior toward one another than toward females. Sugiyama (1968:236) noted:

Males may have a rather strong social bond toward each other. They frequently move in all-male parties, and greater social interaction was observed between them, but was rarely seen between females.

Parties made up of adult females and immatures are a second common pattern. The phenomenon of males preferring other males as companions over anestrous females is expected within a kin-selected, male-retentive, territorial system as hypothesized in the introduction but why did females prefer one another?

There are several possible explanations: (1) Mothers carrying infants are under a double metabolic stress in providing calories for their infants through milk production (Gunther 1971) and in burning extra calories by transporting their infants (Taylor et al. 1970). Such females should tend to range the minimum distance in a day that takes them to adequate food patches. Their metabolic needs and concomitant disinclination to travel long distances may make them incompatible companions for adult males and/or adult females without infants (see also Rodman 1973). (2) Females are of lower status than adult males and may be cheated by them during mutual grooming (see below). (3) Females who elect to travel with males, who characteristically spend less time resting, would have less leisure time than otherwise, a situation possibly disadvantageous for a female with dependent offspring. (4) If males occasionally engage in territorial clashes, a female who travels with them may be endangering herself unnecessarily. (5) In the interest of the social development of her growing offspring, a female should associate with other mothers so that their offspring can play together and develop social identities. This may be especially important for young males, if the social system is male-retentive, because they will be mutually dependent in adulthood. (6) Some adult females may be related to one another as closely as are the males: either because a young female never emigrated and continues to socialize with her mother or because some females at Ngogo emigrated to it but were born and raised together in the same neighboring community. I suspect that all of these factors may influence the females' choice of companions, but this choice may occur by default: (7) because males do not favor anestrous females as companions, they are left only with one another.

Traveling companions of estrous females. I did not often see adult females with what I judged to be full perineal swellings. Perineal swellings were judged maximal by the same criteria as used by McGinnis (1979), but, as discussed below, many such swellings on young females may be anestrous. When equating maximal swelling with sexual receptivity and estrus, it seems logical that such females travel with one or more adult males in preference to other females, unless mating with such males (i.e., of the female's natal community) is not in her best reproductive interests. But adult females of all classes were companions 38 percent more often than were adult and subadult males (table 27). Even so, this difference among estrous females is much more skewed toward males as companions than the overall 98 percent difference reflecting the preference of adult females without infants for other females (table 24). This suggests a tendency for estrous females to travel with males more often than they otherwise would have. Of all 91 traveling parties containing estrous females, 34 percent were solitary.

Nishida (1968:185, 191) reported, "estrus [sic] females were most often found in large, mixed groups . . ." and "Chimpanzees tend to form a relatively large congregation when estrous females are involved." Reynolds and Reynolds (1965:418) reported that 74 percent of all groups containing estrous females also contained one or more adult males. They concluded, "This indicates that the mother in estrus tends to leave her group to join males, or that males join up with mother bands when one of them is in estrus."

During this study I saw none of the eight adult females accompanied by immature offspring develop a perineal swelling, and I saw only one estrous female with a very large infant (or small juvenile?) at Kanyawara. Usually estrous females were young and without offspring. In general, adult males paid little or no overt attention to this class of female. Perhaps this was because most of these females may not have been fertile, or in real estrus. Short (1979) estimated that about half of a wild female's sexual swellings would be anestrous and occur during her immature stage prior to her first pregnancy. Owl, a young or possibly subadult female, for instance, appeared to remain in full swelling for up to a week at a time yet females her age apparently are infertile despite perineal tumescence (Asdell 1946; Goodall 1968). Membership of parties containing a female with a large perineal swelling in Kibale and lack of sexual behavior in the presence of such females suggest that males knew when it was worthwhile to attempt to mate with such females. How they knew that some females were not ovulating is an interesting question. The section on mating behavior discusses this in more detail. In general, the tendency was for females with large perineal swellings to travel more often with males than did females without swellings.

TABLE 27. TRAVELING COMPANIONS OF ADULT FEMALE CHIMPANZEES WITH MAXIMAL* PERINEAL SWELLINGS IN KIBALE FOREST, UGANDA, BETWEEN DECEMBER 1976 AND MAY 1978 AND JANUARY AND MAY 1981.

Adult female with maximal[a] perineal swelling	In traveling party (N = 91) with one or more								
	Solitary	Adult male	Subadult male	Juvenile male	Adult female without perineal swelling	Adult female with perineal swelling	Adult female with infant	Juvenile female	Total companions
Number of times seen traveling	31	43	15	17	45	13	22	14	169

[a] Perineal swellings were judged "maximal" by estimation in field only.

Associations between individuals

Associations between chimpanzees occurred even when the individuals had not traveled together. Aggregations at feeding sites often were much larger than the traveling parties that contributed to them, and peaceful interactions occurred in these aggregations between individuals who arrived and departed the aggregation in different parties. Membership of these temporary associations is a further indicator of the social system of chimpanzees.

Table 28 summarizes the associations I observed between all chimpanzees past infancy whom I recognized at Ngogo. Note that no single individual was seen with all other known individuals. I saw both Raw Patch and Silverback with 25 of the 38 individuals past infancy whom I recognized at Ngogo. Of these 25 chimpanzees only 19 associated with both adult males, so that 31 of the 38 recognizables at Ngogo associated with one of these two males, who often associated with one another. Other adult males who associated with Raw Patch and Silverback also associated with another four individuals. This social network provides the working definition of the Ngogo community. Only Kella, Polly, and Punch were not seen in association with the Ngogo males, but I saw these females too few times to be able to conclude their membership in the same community.

Because males in table 28 held the four top-ranking positions for the greatest number of different individuals with whom they associated, I tested the separate classes of males and females for differences in sizes of their social networks. Males associated with 1.03 different individuals per sighting, on average, compared with 0.93 different individuals per sighting by the females. Although suggestive, this difference is not significant (Wilcoxon 2 sample rank test, $T_2 = 316$, d.f. $= 13, 17, p > 0.05$).

The raw data in table 28 do not convey at a glance a firm idea of the tendency of each sex to associate more with individuals of like sex than of unlike sex. By computing an index of familiarity between individuals for all observed dyadic associations (often within larger aggregations) this tendency emerges. The index of familiarity used (after Nishida 1968) is as follows: % familiarity $= (c/a + b + c)$ 100, where a is the number of associations containing individual a, but not b; b is the number of associations containing individual b, but not a; and c is the number of associations containing both individuals a and b.

Using data from table 28, table 29 presents indexes of familiarity for all pairs of adults I observed in association at Ngogo. Because a large proportion of my sightings were of solitary apes and because most traveling parties and other associations were small, most of the indexes listed are low. The pair exhibiting the highest index, Gray and Zira, traveled together so frequently that I was surprised when I found one without the other, yet the index of familiarity for these two is only 51 percent.

Table 29 is still not adequate for assessing sexual differences at a glance. By computing a mean index of familiarity for all observed pairings within the three classes, male-male, male-female, and female-female, differences finally emerge. The mean index for all-male pairings is 3.9 percent; that for male-female pairings is 2.5 percent; and the index for all-female pairings is 4.1 percent (females Polly and Punch were not included in the computation of indexes). The salient aspects of this comparison are: (1) adult males formed social bonds that were more often reinforced among themselves than among females; (2) adult females formed social bonds that were more often reinforced among themselves than among males; (3) social bonds between males and females were the weakest and least often reinforced.

Further examples of these tendencies are illustrated in table 30, which lists all traveling companions of five of the seven adult chimpanzees whom I contacted most frequently at Ngogo. The top-ranking 2, 4, and 5 traveling companions of females Blondie, Gray, and Owl, respectively all were females. The top-ranking 5 and 4 traveling companions of males Raw Patch and Stump, respectively, all were males. Table 30 gives an indication of the fidelity of companions. It is difficult for me not to think of Gray and Zira, Raw Patch and Stump, and Stump and Silverback as friends, or not to classify Owl with Gray and Zira, or Blondie with Ardith or La, as preferential companions. The affiliative behavior of these pairs included long sessions of mutual grooming that reinforced my impression that the social bonds between them were strong, even though they did not always travel together.

After a similar analysis of interindividual affinities among chimpanzees in Mahale Mountains, Nishida (1968) made the following clarifications: "(i) social bonds are stronger among adult males

TABLE 28. INTERINDIVIDUAL ASSOCIATIONS BETWEEN RECOGNIZABLE, POST-INFANT-AGED CHIMPANZEES SEEN TOGETHER, BUT NOT NECESSARILY TRAVELING TOGETHER AT NGOGO, KIBALE FOREST, UGANDA, BETWEEN JANUARY 1977 AND MAY 1978.

Name	Age-sex class	No. of times seen	No. of post infant-aged individuals associated with	Ardith	Ashly	Blondie	Bess	B.U.L.B.	Clovis	Clark	Dumbo	Eagle	Eskimo	Farkle	Fearless	Felony	Fern	Gray	Hump	Ita	Joe	Kella	La	Mom	Mac	Nane	Newman	Notches	Owl	Polly	Punch	Phantom	Quilla	Raw Patch	Satan	Shemp	Silverback	Spots	Stump	Zira	Zane
Ardith	AF w/IM	12	16																																							
Ashly	JM	18	22																																							
Blondie	AF w/IM	28	18	6																																					
Bess	JF	29	20	3	1	27																																			
B.U.L.B.	AF	6	3																																							
Clovis	AF w/IM	8	10	1	2	2	2																																		
Clark	JM	7	3						5																																	
Dumbo	AF	9	8																																							
Eagle	AM	4	9		1																																				
Eskimo	AM	8	12								3																															
Farkle	AF w/IF	9	14	1	1																																				
Fearless	SAM	12	16	1	1	1								2																											
Felony	JF	11	16	1	1	2								9	3																										

Name	Class																								
Fern	JF	13	15			1 1						1 1 3 9														
Gray	AF	35	20	2 1		3		7			3	3													
Hump	AF	8	8	1		1		1																	
Ita	AF	17	22	4 1 1	2	2			2 3 3 2 3																
Joe	AM	1	2																						
Kella	AF w/IM	3	3	3 3					1 5																
La	AF w/IF	9	18	3 4 5	2			2			1														
Mom	AF w/I?	4	12	1		1 1		1 1			6 1														
Mac	JM	4	12	1	1 1	1 1		1 1			1 1	1 4													
Nane	AM	5	3					1																	
Newman	AM	10	24	1 1 1 1		3 1 5	1	1 1 1 4		1 1 1															
Notches	AM	5	4	1 1 1	1			1 1																	
Owl	AF	39	20	2 1 3 4		1		5 2 1 6	1 5	3	3														
Polly	AF w/I?	1	1					1																	
Punch	AF	1	1																						
Phantom	JM	8	5	3		1 1	1	1		1	1 2														
Quilla	AF	4	9	1	1	1 1		2	1	2	3 6 1 4	3													
Raw Patch	AM	60	25	1 6 4 4 1	2 2	4 1 3		9 4 4 1		2	1														
Satan	AM	3	3																						
Shemp	AM	6	14	1 2	1	3 3		2 1 1 6	1	3 1 1	3 5	1 1 1													
Silverback	AM	25	25	1 2 3 1	1 2	3 4 2		4 3 2 6 1 6	3 1	1	5	5 2													
Spots	AM	7	14	1 3 1 1	1 1	1		1 1			1	1													
Stump	AM	23	19	1 2		3 1 3 2	1	3 2 2 2 2	2	1 1 1 4	3 2 2	1 5 8	2 9	3											
Zira	AF	24	17	2 2 1		6 3	1	20 6	6	3 4	4 6	1 6	7	1											
Zane	JF	7	6	1		6	1	6		3	3	1		4											

	Eagle	Eskimo	Fearless	Joe	Nane	Newman	Notches	Raw Patch	Satan	Shemp	Silverback	Spots	Stump	Ardith	Blondie	B.U.L.B	Clovis	Dumbo	Farkle	Gray Hump	Itta	Kella	La Mom	Owl	Polly	Punch	Quilla	Zira
MALES:																												
Eagle	–																											
Eskimo	7	–																										
Fearless	–	–	–																									
Joe	–	–	–	–																								
Nane	8	38	5	10	–																							
Newman	–	–	–	–	–	–																						
Notches	–	5	1	2	5	9	–																					
Raw Patch	–	7	–	–	–	23	2	–																				
Satan	–	–	–	–	–	17	–	–	–																			
Shemp	11	–	12	–	–	–	–	2	11	–																		
Silverback	12	14	12	–	–	14	–	6	–	7	–																	
Spots	–	–	–	–	4	–	–	8	–	–	3	–																
Stump	4	11	9	–	4	–	–	11	–	7	23	–	–															

FEMALES:

Ardith
Blondie
B.U.L.B.
Clovis
Dumbo
Farkle
Gray
Hump
Ita
Kella
La
Mom
Owl
Polly
Punch
Quilla
Zira

TABLE 30. SUMMARY OF INTERINDIVIDUAL AFFINITIES OF 5 OF THE 7 MOST FREQUENTLY OBSERVED CHIMPANZEES AT NGOGO, KIBALE NATURE RESERVE, UGANDA.

AF, Blondie, with IM, Butch, and JF, Bess		AF, Gray		AF, Owl		AM, Raw Patch		AM, Stump	
Companion	N	Companion	N	Companion	N	Companion	N	Companion	N
AF, Ardith, and	5	AF, Zira	29	AF, Gray	5	AM, Stump	8	AM, Silverback	9
IM, Anson	(2)	JF, Zane	8	AF, Zira	5	JM, Ashly	5	AM, Raw Patch	8
AF, La, with	4	AF, Owl	5	AF, Ita	4	AM, Nane	5	AM, Newman	3
IF, Lysa		AF, Ita	3	JF, Bess	3	AM, Newman	5	SAM, Fearless	2
AM, Raw Patch	4	AM, Raw Patch	3	AF, Blondie, with	2	AM, Spots	5	AF, Ita	2
AF, Clovis, with	2	AF, Blondie, with	2	IM, Butch		AF, Blondie, with	4	AF, Dumbo	1
IM, Chita		IM, Butch, and		SAM, Fearless	2	IM, Butch, and		AM, Eskimo	1
AF, Farkle, with	2	JF, Bess	2	AF, La, With	2	JF, Bess	(3)	AF, Gray	1
IF, Fanny,		IF, La, With		IF, Lysa		AF, Hump	4	AM, Nane	1
JF, Fern, and		IF, Lysa	2	AF, Quilla	2	AF, Gray	3	AF, Owl	1
JF, Felony	2	AM, Silverback		JF, Zane	1	AM, Silverback	3	AF, Zira	1
AF, Gray	2	AF, Dumbo	2	AF, Ardith	1	AF, Zira	3		
AF, Ita	2	AM, Eskimo	2	AF, Hump	1	JM, Phantom	2		
AF, Owl	2	AM, Newman	2	AM, Newman	1	AF, B.U.L.B.	1		
JM, Ashly	1	AF, Quilla	1	AM, Stump	1	AF, Dumbo	1		
AF, Hump	1	AM, Spots	1			SAM, Fearless	1		
AM, Newman	1	AM, Stump	1			AF, Ita	1		
AM, Notches	1					AM, Joe	1		
AM, Spots	1					AF, La, with	1		
AF, Zira	1					IF, Lysa			
						AM, Notches	1		
						AM, Shemp	1		
Total individuals:	20		17		15		22		11
Total males:	7		7		4		12		6
Total females:	13		10		11		10		5

NOTE: Data recorded ...

"(ii) male-female familiarity is stronger than female-female familiarity."

Nishida concluded, "adult males may play the most important role in unity of an entire group." Nishida (1979) later quantified, "male cohesiveness was roughly twice that of females." A striking difference between his findings and mine is the fidelity of Mahale females toward adult males rather than other adult females, contrasted with the opposite tendency of the Ngogo females of this study. The tendency of the Ngogo females was further demonstrated by their patterns of social grooming. Nishida (1979) noted that even within presumed matrilines affinities were weaker than between males and females. No close combinations of an old female with a young, mature one (such as Gray and Zira at Ngogo) were seen at Mahale. An explanation for the differences observed in the females of these two areas is lacking, but it is noteworthy that the adult males of Ngogo and Mahale both were characterized by strong social bonds among themselves as predicted by the hypothesis of sexual selection and territoriality.

SOCIAL GROOMING

Social grooming was a recurrent interaction among chimpanzees in Kibale Forest. Social grooming is defined here as the fine manipulation of the body surface of one chimpanzee by a second for the apparent primary purpose of removing foreign matter, ectoparasites, or dead skin, etc. from the skin of the first chimpanzee. During this process the groomer used one or both hands, lips, and/or tongue for manipulating the skin and hair of the groomee, for removal of foreign matter, or for adjusting the attitude of a body part. Often the groomer visually scrutinized, from approximately 15 cm or even closer, the area being groomed. Sessions between a single pair lasted up to 2 hours, including brief lulls, with groomer–groomee rolls switching up to 40 times during a session. Lip-smacking and/or popping sounds were often made by the groomer, while the groomee usually remained quiet and immobile unless it was grooming a third chimpanzee, idlely grooming itself, or shifting position to present a new part of its anatomy to the groomer.

Social grooming was a quiet, relaxed interaction which usually occurred during periods when social or environmental tensions appeared

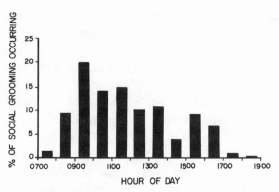

FIGURE 16. Hourly distribution of time spent in social grooming by chimpanzees in Kibale Forest, Uganda between December 1976–May 1978.

minimal. Sessions were initiated most often during or after the latter stage of intensive feeding. Figure 16 illustrates the diurnal distirbution of time spent in social grooming sessions by Kibale chimpanzees. Mid to late morning was the prime time for grooming, following the early morning feeding peak. After 1700 hours I rarely saw social grooming. Goodall (1968), Sugiyama (1969) and Nishida (1970) reported brief sessions of social grooming occurring in times of social stress as an appeasement, reassurance, or displacement behavior. I did not observe grooming in these contexts, perhaps because I rarely observed social stress.

Normally chimpanzees did not maintain contact proximity during feeding or nongrooming periods of rest. For social grooming to occur one ape had to close the gap between it and its potential partner. Social grooming was initiated by one chimpanzee walking or climbing to a second, often lip-smacking during the approach, presumably to signal its intentions, and, then, by either beginning to groom the second or by presenting some part of its body to the second to be groomed.

Once initiated, social grooming took on overtones of a dynamic but friendly contest over who would be groomed the most. The immediate benefits of being groomed apparently outweighed the potential and less gratifying benefits of grooming. As a groomer warmed to its task, the groomee serially presented a selection of body parts in need of grooming—until the groomer stopped and presented a part of its body to the groomee. Roles then reversed; the new groomee was never at a loss in presenting successive body parts in need of grooming. Usually roles reversed repeatedly during sessions of moderate to great length. Occasionally an initial groomee failed to reciprocate, and the session ended earlier than normal, with the groomer having not been groomed. My impression was that nonreciprocal sessions between mature individuals occurred because the initiator groomed its partner too early in the latter's feeding bout, or groomed an individual of much higher social rank, or both.

Once under way social grooming seemed to be a mutually enjoyable experience. Personal space and normal body protocols broke down during sessions. A groomer was at liberty to grasp the head of the groomee and gently force it into an attitude that facilitated grooming. With apparent trust and vulnerability a groomee would lie supine with both legs extended upward so that a partner could groom its groin, or a groomee would lie prone, draped along a limb with arms and legs dangling downward into space, so that a partner could groom its lower dorsum or perineum.

Termination usually occurred when a groomee failed to switch roles and reciprocate when the groomer stopped and presented to it to be groomed. Occasionally, an original groomer regroomed its reluctant partner for several minutes longer, then stopped and waited again to be groomed. Sometimes a patient former groomer eventually was groomed by a reluctant partner, who seemed to do so only because the groomer would not move away. I grew to think of these lulls as "grooming stand offs," when one partner patently desired a continued session while the other showed no inclination to continue as groomer but did not object to being groomed. The intent not to reciprocate, responsible for termination of a session, often was signalled by the last groomee getting up and walking away or staying and beginning to feed or self-groom.

Some stereotypic modes or postures of social grooming varied between communities of chimpanzees. For example, during 38 percent of all nonmaternal sessions of social grooming by the Kanyawara chimpanzees in 1981, one individual, usually the groomer, grasped the hand of its partner and held it aloft. The groomer then had only one hand available for grooming. Although the A-frame, hand-clasp posture facilitated grooming of the underarm region of the groomee, sometimes that region was not a grooming target during the session. Neither did the hand clasp seem to lend stability to the pair. The hand clasp appeared much like a social convention among the Kanyawara chimpanzees. It did not occur at all among the apes of Ngogo. McGrew and Tutin (1978) reported the identical phenomenon among the chimpanzees of Mahale Mountains and noted its absence among the chimpanzees of Gombe 170 km north (only about 50 km separate the nearest portions of the two populations' respective ranges). McGrew and Tutin evaluated this difference between populations and concluded it was a valid example of *cultural* processes among chimpanzees, albeit one with an unknown ontogeny. This community-specific difference exhibited by three separate populations illustrates the complexity of the social grooming relationship among chimpanzees.

Functions of social grooming

Because of the physical contact and trust involved, Goodall (1968:268) suggested that social grooming among chimpanzees "may be vital in helping to maintain reasonably equable social relationships amongst these easily aroused higher primates." Sparks (1967) interpreted grooming as social lubricant to ease the passage through interactions involving individuals of unequal status in a social system characterized by a rigid dominance hierarchy. Sugiyama (1969) tentatively implied that Budongo chimpanzees selectively groomed more dominant individuals for the same reason. But he also stated, "Grooming was frequently observed after the rain stopped and must have occurred mainly due to hygienic requests." None of the above workers presented a data-based test of their ideas.

Sayfarth (1976) reported that among savanna baboons (*Papio cynocephalus*) in Amboseli an individual was most likely to receive grooming if it was: (1) female, (2) high ranking, and (3) lactating to nurse a young infant. Changes in the latter two conditions altered the frequency at which solicitations for grooming were successful. Differences in the proximal causes of social grooming among chimpanzees have been little explored. What are the main functions of social grooming? My impression is that the brief sessions occurring in times of social stress were exceptions to the basic phenomenon of social grooming; they seem to be a ritualized appeasement, a side product of social grooming per se. Initially I suspected that chimpanzees who groomed one another reciprocally were engaged primarily in an exchange service, i.e., grooming an individual's inaccessible regions in trade for that individual's grooming their own.

The cost–benefit ratio could be analyzed as a version of Trivers' (1971) model of reciprocal altruism, without appreciable lag time for reciprocation. If chimpanzees were grooming each other randomly with respect to coverage of body areas of a particular partner, then appeasement and/or grooming as social facilitation makes sense as an explanation. If, however, chimpanzees groomed one another predominantly at areas of the body that were not susceptible to efficient self-grooming, the obvious functional explanation probably is the correct one.

Table 31 summarizes the target areas of the body allogroomed by chimpanzees in 84 sessions of nonmaternal social grooming between December 1976 and May 1978. Each time a chimpanzee began grooming or switched to a new target area of the groomee I recorded that new area in my notes. Ideally the transition times should have been recorded as well, so that the total duration of grooming time that each part of the body received could be determined. In practice I found that I could not keep an accurate account of transition times on a continuous basis during grooming sessions lasting more than a half hour, so table 31 is necessarily a summation of how many times an area of the body received allogrooming attention, regardless of its duration.

The great majority of areas allogroomed were not visible to their owner. Only 4.5 percent of the areas receiving social grooming were areas that could be groomed as easily by the groomee as by the groomer, yet these areas accounted for approximately 55 percent of the chimpanzee's total body area (extrapolation from The Committee on Injuries 1971:139; also see Hutchins and Barash 1976:146). In contrast, over 61 percent of allogrooming targets (head, neck, back and sides, for instance) represented only 24 percent of chimpanzee body area, but those areas were not visible to self-grooming.

Social-grooming technique involves intense, short-distance scanning of the area being groomed, which suggests that thorough self-grooming is impossible. It is notable here that the chimpanzee louse (*Pedicularis schaefi*) is so well adapted to its host that it immobilizes when exposed to strong light (Kuhn 1968), as normally happens when hairs are parted during grooming, and thus becomes difficult to detect. Clearly the apparently exaggerated close attention chimpanzees give to their grooming is in fact not exaggerated at all, but is necessary for effective removal of these ectoparasites. Chimpanzees who groomed one another concentrated

TABLE 31. SUMMATION OF TIMES THAT BODY AREAS WERE ALLOGROOMED DURING SOCIAL GROOMING SESSIONS BETWEEN CHIMPANZEES (EXCLUDING MOTHER-INFANT PAIRS) IN KIBALE FOREST, UGANDA.

Accessibility to self grooming	Area of body	Proportion of body surface (%)	No. of times allogroomed	Proportion of total times allogroomed (%)	Summation of proportions allogroomed (%)
Cannot be seen	head or neck	9	92	24.3	
	back or sides	15	141	37.2	
	armpits	2	21	5.5	82.8
	perineum	1	24	6.3	
	backs of thighs or arms	12	36	9.5	
Poorly visible and/or accessible to one hand only	shoulders	4	28	7.4	12.7
	groin	2	20	5.3	
Readily accessible	chest or abdomen	15	5	1.3	
	arms or hands	40	9	2.4	4.5
	legs or feet		3	0.8	
Total		100	379	100	

on body areas that were not susceptible to self-grooming (often at the direction of the groomee), and by doing so performed a valuable service that usually was repaid in kind.

Goodall (1968:263) reported that among Gombe chimpanzees, "the head, back, and rump were the most frequently presented" for grooming. Hutchins and Barash (1976) monitored social grooming among captive lion-tailed macaques (*Macaca silenus*), ring-tailed lemurs (*Lemur catta*), and Celebes macaques (*Macaca niger*) and found that social grooming was directed with statisically greater frequency toward areas of the groomees' bodies inaccessible to self-grooming.

Reports on the consequences of not being groomed are rare in the literature. Struhsaker (1967) reported a vervet monkey (*Cerocopithecus aethiops*) who acquired a heavy infestation of ticks during an absence from his group. Washburn and DeVore (1961) reported group-living baboons as being relatively tick-free compared to the heavy infestations harbored by nonsocially grooming mammals, lions and dogs, in the same habitat. Ruch (1959) noted that solitary male baboons were heavily infested with ticks. Smith (1977) reported that infestations of botfly larvae among howling monkeys (*Alouatta palliata*) were sometimes so heavy as to cause or hasten mortality; among these monkeys grooming was a rare event. Botfly larvae infest primates in Kibale Forest too (personal observation during *post mortum* of a juvenile red colobus with a larva in his foot), but grooming seems common among Kibale primates. Social grooming probably ameliorates botfly infestations as well as external parasitic infections. Almost certainly the consequences of not being groomed are poorer health, which, if severe, will lead to reduced reproductive success.

Who grooms whom

Because social grooming is of benefit to the groomee and is performed at some cost to the groomer, the theory of natural selection predicts that individual chimpanzees should distribute their grooming services to individuals to whom they are most closely related, so as to benefit by increased individual and/or inclusive fitness, and/or to individuals who are most likely to reciprocate their grooming favors. As predicted, grooming interactions between chimpanzees in Kibale Forest were not randomly distributed.

Relative frequency. Table 32 summarizes social grooming sessions by Kibale chimpanzees. Adult females with infants groomed their own offspring significantly more frequently (considering each session as a single grooming event) than they groomed other individuals, as compared to females without infants, who groomed other adults more frequently than they groomed immature chimpanzees ($\chi^2 = 29.067$, d.f. = 1, $p \leqslant 0.0005$). Excluding observations of maternal grooming and considering sex differences only in sessions of social grooming between adults and other adults or juveniles (not suspected to be offspring of the grooming partner) a pattern of unisexual partnerships emerges. Females groomed other females significantly more frequently than they groomed males, and males groomed other males more frequently than they groomed females ($\chi^2 = 32.56$, d.f. = 1, $p \leqslant 0.0005$). I observed partnerships between males the most frequently (considering nonmaternal sessions) despite females having been better represented in my total observation time.

Goodall (1968:265) reported that partner selection for social grooming at Gombe depended upon dominance ranks, estrous condition of females, and individual preferences. She found that males groomed males more often than other classes; females with infants groomed their own offspring most, then high-ranking adult males, then other females; and females without infants groomed one another most frequently. These data agree with mine. Simpson (1973) reported that males at Gombe groomed one another most frequently, as did Sugiyama (1969), although the latter also found that females groomed males more than other females. Van Hooff (1973) reported that males in a semicaptive group of chimpanzees groomed and were groomed frequently; the group's females, despite having groomed often, were groomed little. Mason (1970) reported that captive female chimpanzees tended to groom a more dominant partner. In contrast, Reynolds and Reynolds (1965) found that females groomed other females most often (as in the present study), but unlike Kibale chimpanzees, so did males more than they groomed one another. These data may be biased, though, because female partners of males were in estrus during 14 of the 21 observations. Kano (1980) reported that of 155 sessions of social grooming among pygmy chimpanzees at Wamba, 54.2 percent occurred between males and females (none of which occurred in the context of courtship), 33.5 occurred between females, and only 12.3 occurred between males. But, unlike common chimpanzees, agonistic interactions

TABLE 32. SUMMARY OF COMPLETE SOCIAL GROOMING SESSIONS BY CHIMPANZEES OBSERVED IN KIBALE FOREST, UGANDA, BETWEEN DECEMBER 1976–MAY 1978 AND JANUARY–MAY 1981.

Grooming pair	No. of sessions observed	Reciprocity			Total time spent grooming minus lulls (min)	Duration of grooming session (min)		
		yes	no	?		$\bar{x}$	Range	S.D.
Adult female-immature offspring	27	10	16	1	281.78	10.83	0.09–70.0	15.70
Immature sibling-immature sibling	3	–	3	–	8.50	2.83	1.0–6.0	2.77
Shared matriline not assumed:								
Female-female	26	18	7	1	403.00	15.50	1.0–101.0	21.83
Female-male	24	13	9	2	245.83	10.40	0.33–29.0	8.89
Male-male	28	21	7	–	497.50	17.76	1.0–78.0	16.97
Total:	108	62	42	4	1,436.61	13.42[a]	0.09–101.0[a]	16.44[a]

[a] Average.

between males were common and affiliative behaviors were rare. Maternal grooming sessions were more frequent than any type of session between adults.

The general pattern most supported by this study and the above workers is for grooming partnerships to form primarily of like-sexed common chimpanzees and for mothers to direct more grooming toward their offspring than toward other individuals.

Duration. Grooming sessions between males in Kibale lasted longer ($\bar{x}$ = 17.76 min) than sessions between all other age-sex-reproductive classes. Durations of sessions reported here are summations of time actually spent grooming *not* including time spent in lulls. Of all class combinations including one or more adults as partners, grooming sessions that included a male and female lasted the lowest mean duration (see table 32). Nonmaternal sessions between like-sexed partners lasted for similar durations and were longer than sessions between a male and female (Wilcoxon rank sum test, $T_2, = 539, n_1 = 28, n_2 = 24, p \leq 0.05$ for all-male sessions compared to mixed sessions: $T_2 = 601, n_1 = 22, n_2 = 26, p > 0.05$, for all-female sessions compared to mixed sessions). Although unequal sexual representations in aggregations of chimpanzees may be responsible in part for unequal sexual composition in all nonmaternal sessions when summarized, the tendency for *shorter* bisexual sessions are more independent of party compositions and likely arose from preferences of the animals themselves.

Goodall (1968) did not report durations of grooming sessions on the basis of sexual composition of partnerships, but she did report that sessions lasted up to 2 hours prior to the program of provisioning with bananas and usually were under 1 hour in duration after provisioning had begun. Simpson (1973), reporting on the patterns of social grooming by Gombe males, stated that males of lower status groomed longer; males of higher status groomed more frequently, received grooming more often, and preferred one another to juniors. Individuals were predictable in given pairs, and the longer a male groomed, the more he received grooming. Bauer (1979) examined social grooming in the context of reunions between Gombe chimpanzees. He found that grooming sessions more frequently included an individual if he had given a charging display during the reunion than if he had not, but that sessions lasted longer if no charging display had occurred at all. Interestingly, Bauer also reported that grooming partnerships were more likely to be composed of individuals from the two traveling parties who had just met than of two individuals traveling in the same party. Bauer's report indicates that social grooming has an affiliative as well as hygienic function.

Reciprocity. Not all sessions of social grooming were sessions of mutual grooming; reciprocity was unequal between age-sex classes (table 32). The lowest frequency of reciprocation occurred between mothers and their immature offspring and between sibling pairs which included an infant. The low rate of reciprocation was due to the infants' failure to groom their mothers or siblings. Among pairs of groomers not assumed to be related matrilineally, frequency of reciprocation was higher in like-sexed pairs (72% for female-female pairs, 75% for male-male pairs) than for pairs of mixed sexes (59%). In six of the nine instances of nonreciprocity an adult male failed to groom a female who had groomed him. In the remaining three instances a single young adult male, Raw Patch, initiated sessions with a female who did not reciprocate. For 17 months Raw Patch had an unhealed wound approximately 40 cm^2 in area on his lower back, a wound he was unable to inspect visually. He had a habit of frequently touching the periphery of the wound, as if it were itchy and he was reluctant to scratch the wound itself. My impression was that he initiated sessions overeagerly, possibly spurred by his wound, during periods when his partner was more interested in foraging than in grooming. In general, males seemed more likely to be nonreciprocators toward females than vice versa, although the actual proportion of grooming time that males were groomees during sessions of mixed sexes (53%) differed little from the proportion when females were groomees (47%).

The nonrandom pattern of reciprocity in nonmaternal sessions of grooming suggest that social bonds are stronger between like sexes than between mixed sexes. Goodall (1968:264) noted that nonreciprocity among grooming pairs of Gombe chimpanzees primarily occurred after adolescent apes groomed older males. Our observations agree, and both sets of data suggest that dominance status as well as sex influence reciprocity.

Reciprocity was not always clear cut. It was apparent that most groomers wanted to be groomed after they had groomed, and my impression was that the initial groomees knew this. Individuals of lower social status (not considering infants) rarely were nonreciprocators and rarely terminated a session directly. But some became functional nonreciprocators while putting on a mere appearance of reciprocation. For example, during a session lasting at least 70 min between an old female, Farkle, and her approximately 9-year-old daughter, Felony, the latter appeared to tire of reciprocating. Each time it came her turn to groom she sat behind her mother and, while gazing about the tree crown (for fruit?) and at her two younger sisters, Felony ran her hand across her mother's back and parted the hairs as if grooming, but she did not look at the skin or, in fact, do any real grooming at all. After each sham reciprocation Farkle meticulously groomed her daughter.

Significance of social grooming

Social grooming is not a simple phenomenon among chimpanzees. The primary function of social grooming seems to be hygienic, but the sociosexual differences in the patterns of relative frequencies, durations, and reciprocity of sessions of social grooming suggest that there exist purely social, nonhygienic aspects which influence the expression of grooming behavior. I found only indirect evidence in the literature that deprivation of social grooming significantly alters an individual's fitness, although it seems probable that such is the case. In general, it seemed that each ape was willing to receive grooming but was not always willing to reciprocate. Mothers concentrated their grooming on their immature offspring, a sensible allocation from the perspective of natural selection (Dawkins 1976). Other adults concentrated their grooming on like-sexed individuals. Within the theoretical framework of the hypothesis on sexual selection and territoriality males should prefer other males as grooming partners because they are likely to be more closely related to one another and mutual grooming may enhance their inclusive fitness and because mutual grooming apparently acts to reaffirm bonds between individuals who have been separated (Bauer 1979). Strong social bonds between males may be important in communal defense of their territory.

Females' propensity to groom other females is less easily explained. If females are less likely to be groomed reciprocally by adult males because of differences in dominance status, and if they do not travel with offspring old enough to reciprocate effectively, another female of the same category who may be willing to reciprocate in the sense of Trivers' (1971) reciprocal altruism seems to be the best choice of grooming partners. Pragmatically, grooming partners and traveling companions should be the same individuals, though feeding ecology may dictate against this at times. Females who travel with other females probably groom with them more often for this reason. Another factor which may influence female choice of grooming partners is the degree of relatedness between them. Mother–daughter pairs or sibling females would be more likely to groom one another for reasons related to inclusive fitness and possibly to lifelong affiliations.

Reciprocal social grooming, as a hygienic exchange service and reinforcer of social bonds, appeared to be an important and intrinsically rewarding interaction between individuals. The nonrandom patterns of this exchange indicate individual preferences as the product of natural selection.

DOMINANCE INTERACTIONS

Wilson (1975:287) wrote, "In the language of sociobiology, to dominate is to possess priority of access to the necessities of life and reproduction." Among Kibale chimpanzees I rarely observed overtly competitive interactions for limited resources, indicating clearcut dominance relations between individuals. Those dominant–subordinate interactions I did observe were mainly intrasexual, uncontested, approach-displacement sequences (table 33). Of 27 dominance displacements 8 occurred during foraging and resulted in the more dominant individual gaining preferential access to a small part of a food patch, the whole of which was indefensible. Three other displacements by adult males were directed at temporary near neighbors of an estrous female; the dominant males subsequently mated with the female. Other dominance interactions were less clear cut; some may have resulted in preferential access to food, though this was not possible to ascertain. Adult males were the most dominant individuals and subadult males were their most frequent subordinate targets. The three adult male–adult female interactions consisted of (1) a female who moved horizontally out of an egress route to a food tree so that the adult male climbing behind her could pass, (2) a female who stopped foraging and sat still for 1 min as a young adult male foraged within 1 m of her (as he moved away she resumed foraging), and (3) an estrous female who moved away from the rushing approach of a male who apparently was displacing her two male companions. All dominance

TABLE 33. SUMMARY OF DOMINANCE INTERACTIONS BETWEEN CHIMPANZEES IN KIBALE FOREST, UGANDA.

Dominant individual	Subordinate individual						
	Adult male	Subadult male	Juvenile male	Adult female	Juvenile female	Juvenile sex?	Total
Adult male	RA–D, 2 A–D, 2 A–D, 2	RA–D, 5 RA–S, 1 A–D, 1	A–D, 1	RA–D, 1 A–D, 1 A^a, 1	–	RA–D, 2	19
Adult female	–	–	–	RA–D, 1	RA–D, 1 A–S, 2 D^b, 1	RA–D, 2 A–S, 1	8

NOTE: Key to symbols: RA, rapid advance; A, advance; D, displacement; S, supplantation.
[a] Adult female ceased feeding and all other activity while adult male foraged within 1 m.
[b] Caused traveling juvenile female to leap perpendicularly to an arboreal path by grunting and kicking a foot in her direction.

interactions occurred without physical contact between individuals; rapid retreats by subordinates were not uncommon. Although few in number, my observations and their trends resemble the patterns among the same age-sex-reproductive classes of chimpanzees at Gombe as reported by Bygott (1979) and at Mahale Mountains as reported by Nishida (1979).

The predominant aspect of social relations between Kibale chimpanzees seemed frictionless and free from hints that a dominance hierarchy existed. Ideally, this is the way an advanced and stable hierarchy should appear, but only when all individuals in it are either satisfied with their respective ranks or sure that it is not the appropriate time to attempt to elevate them.

Popp and De Vore (1979) presented a cost-benefit analysis of aggressive competition in a model of optimal behavior to predict the expression of dominance behavior on the premise that "asymmetry in the cost-benefit functions of competitors is the key to termination of aggressive interactions." Determining the cost of an attempted dominance action and the ultimate cumulative rewards if successful, however, is beyond current field techniques used to record the behavior of chimpanzees. But because the cost of physical combat in dominance interactions can be high, especially among male chimpanzees (Teleki 1973; Bygott 1979; Nishida 1979), dominance hierarchies develop. Individuals in Kibale apparently tended to avoid confrontations with others who were more aggressive and/or physically larger. Popp and De Vore (1979:334) summed up the subject well, "Dominance hierarchies do not exist because they bring harmony and stability to the social group, but as the consequence of self-interested action, in the evolutionary sense, by each group member."

SEXUAL BEHAVIOR

Female chimpanzees mate during estrus, approximately midway through the lunar sexual cycle. As we have seen, most females at Ngogo (10 of 13 or 14) were either lactating or suspected post-menopausal during most of the study. None of them showed perineal swellings or other signs of estrus. The remaining three or four resident females (Ita, Owl, Quilla, and Zira) were young, possibly subadults who exhibited periodic full tumescence and detumescence of their perineums but appeared to be ignored by adult males. At first, when I observed feeding aggregations containing adult males and one or more of these females with full perineal swellings, I suspected that mating might occur once the foragers were satiated. But repeatedly such situations failed to result in mating.

Goodall (1968) observed at Gombe that females developed full-sized perineal swellings 10 months before the onset of menses. Asdell (1946) reported a lag time of 2.33 years between the onset of menses at the age of about 9 years old and subsequent conception among chimpanzees in captivity. These data indicate a three-year period in which a nearly mature female may have maximal perineal swellings but not be fertile. Not surprisingly, Goodall (1968:217) noted, "young females, after their first few sexual swellings sometimes appeared to be unattractive to mature males." It seems likely that the three or four nulliparous Kibale females discussed above were in this same infertile category, and hence were not true adults.

The only Ngogo female I observed to mate was Owl. While foraging in a large *F. dawei* with an adult male, Raw Patch, the latter, while feeding desultorily, casually approached her, moved behind her, mounted, and thrust seven or eight times in less than 5 sec. Owl made three high-pitched screams, then moved away.

I observed an Ngogo female with a partial perineal swelling approach and present to an adult male only once:

18 March 1977, 1457, junction *C/8.5*. Zira, with partial swelling, swings 10 m toward R.P., stops and hangs above his head about 1 m for 5–8 sec. After she stops, R.P. reaches his r. arm up, extends 3 fingers into her vagina with a 1 sec wiggle, then sniffs his fingers and looks uninterested as she swings away.

On three other occasions males approached females who were in apparent estrus but were nonreceptive to the males.

20 March 1977, 1706, *C/8.5*. Stump moves toward an adult female with a full swelling [young] and she begins screaming 1-sec screams as he nears her. As he passes along the limb above her head she cringes, crouching low, while watching him and screaming. He stops 2 m away, sits hunched up, and looks at her. She stops screaming, turns and quickly climbs away. He remains.

18 April 1978, 1604, *C.5/8.5*. Owl is in full swelling, as *usual*. She approaches *F. d. #4*. Ashly [large juvenile male with well developed testicles] moves out from *F. d. #4*, still chewing figs, with penis erect, to meet her. I did not see actual contact but looked up as Owl screamed and rushed away from Ashly. He sat with penis erect. After 30 sec Owl rushed past him about 15 m to sit in center route, then enter *F. d. #4*. Ashly urinates then descends at 1607.

30 April 1978, 1107, 1500 *D.5*. Owl [in full swelling] moves up bole of *F. m.* #2 toward Fearless [a subadult male who arrived 25 min before Owl], who moves 8 m downward to "meet" her. As Owl moves upward, *"hooing,"* as she has been doing for 4 min, Fearless has an erection, sitting. Owl moves toward him, stops, screams, backs down, sits [unable to forage there] for 3 min. After 2 min more she move up to feed within 2 m of him. [He "ignores" her.]

Goodall (1968) also reported that some females fled some males who approached them to mate.

Matings were also rare among the chimpanzees of Kanyawara. During a 3.5 hour midday period in 1981, however, within a party of 10–12 individuals including several adult males plus two adult females with attendant offspring foraging in a *F. natalensis,* I observed several matings. Anka, a middle-aged multiparous female, appeared to be in maximal swelling. She mated 11 times with four or more males, and initiated mating by approaching and presenting her swollen perineum to a male prior to eight of the matings. Actual mounting and thrusting lasted an average of about 6 sec ($R = 1.5$–10 sec). Tutin (1975) reported the average duration of copulations among Gombe chimpanzees as about 8 sec.

Anka's youngest offspring, Auk, a juvenile male 4–5 years old, interfered during nine of the eleven matings. His interference usually consisted of quickly scrambling through the foliage to climb on his mother's back, face the mounted male, and push and tug at the male's abdomen. Auk sometimes screamed while rushing toward the mating pair and continued to scream if the male made an aggressive move toward him. On one occasion a second immature male, Vik, the 3-year-old infant of another female present, rushed to climb on Anka's back along with Auk and grappled with the mounted male. This interference by immatures seemed to hamper, though not completely dissuade, the mating pair. When they separated, Auk sat between them and behaved as though vigilant of a recurrence of mating. Pusey (symposium presentation, 1973) reported such mating interference by immatures at Gombe as common. Interference with mating, if successful, is adaptive from the infant's point of view, because it delays the time of weaning and prolongs the parental investment its mother donates toward it (Trivers 1974). But from the perspective of the mother, who is ready to mate again and whose future reproductive success is dependent upon mating, interference by her youngest offspring is a detriment which she must combat, or at least overrule.

The normal mating pattern among wild chimpanzees has long been understood to be typified by the promiscuity of females in the provisioning areas of Gombe (Goodall 1968) and of Mahale (Nishida 1979). Data collected away from the provisioning areas (McGinnis 1979; Nishida 1979; Pusey 1979) indicate that some females are mated almost exclusively by a single male during their estrous phase, and also that some males are able to defend an estrous female (as a reproductive resource) by taking her on "safari" or by excluding other males by dominance action from mating with her in mixed aggregations. Tutin (1975) reported that about 75 percent of all copulations at Gombe occurred in noncompetitive situations—from a male's point of view. Tutin also suggested that conceptions possibly were more likely during consort safaris than after more promiscuous matings in the provisioning area.

Female discrimination of mating partners during the period in which she is likely to conceive makes sense from the perspective of natural selection theory. But, unless the males present are very similar to one another genetically, female promiscuity during that period does not. But promiscuity is not restricted to artificially provisioned situations. It was reported by Reynolds and Reynolds (1965) for chimpanzees in Budongo Forest, and I observed such matings at Kanyawara. Kano (1980) reported similar promiscuity among pygmy chimpanzees and suggested that a small proportion of total matings seemed aimed *primarily* at inducing a male to share food rather than at simple reproduction. This secondary correlate of promiscuity may be enhanced in a provisioning situation.

A strong influence on the decision of an estrous female over whether to mate with several insistent males in a provisioning area, with whom she might not chose to mate outside it, is the cost of avoiding the provisioning area. In the heyday of provisioning at Gombe and Mahale Mountains a female accustomed to receiving quantities of bananas or sugar cane might have been more likely to heed the needs of her stomach than to be selective about the genes that might combine in her future offspring. This inadvertent "prostitution" among chimpanzees is suggested by the uneven tendency of some females away from provisioning areas to avoid mating with males with whom they apparently do not want to mate and to be monopolized by males who are capable of excluding other males.

Females' choice and female defensibility by males both exact a higher cost than normal in provisioning areas, partly because they consistently attract unnaturally large aggregations of individuals (Wrangham 1974). It seems likely that additional observations of chimpanzees mating outside the context of provisioning areas will reinforce a picture of mating choice rather than wide promiscuity, though promiscuity is patently a natural dimension in the mating system of wild chimpanzees. Again, the factor that would tend to reduce the importance of choice of mates for a female is the likelihood of high relatedness and shared genes between a community's males.

Despite shared genes, males should want to maintain exclusive access to an estrous female. The salient aspect of reproductive behavior among chimpanzees is that, for males, opportunities to mate and reproduce are relatively rare. This is also true for gorillas (Schaller 1963) and orangutans (Rodman 1973a, 1979; MacKinnon 1979). Because opportunities are so rare, social behaviors by some males which make them less rare than for other males will be favored by natural selection. (The role of male reproductive strategies in shaping the social system of chimpanzees is discussed in detail in the next chapter.)

LONG-RANGE CALLING

Chimpanzees exhibit a large vocal repertoire of distinct and intergrading signals (Goodall 1968; Marler 1969). These are unusual among nonhuman primates because of the extensive intergrading of signals among and between vocalizers. One of the most impressive of chimpanzee vocalizations is the long-range "pant-hoot" (Goodall 1968), a four-phased series of clear hoots, wails, shrieks, and roars that rise, build to a climax of high intensity, then trail off into diminishing clear hoots that resemble the initial build up in intensity but in reverse order. These elements occasionally were juxtaposed into an unusual order or uttered in incomplete form. Sometimes a slow build up of rising long hoots or wails drifted back down the scale without having reached an intense climax of rapid hoots and/or a roar. Buttress drumming or slapping sometimes accompanied pant-hooting. Often chimpanzees joined into pant-hooting choruses in which different members specialized on particular sounds; one chimpanzee gave a series of hoots while a second gave rapid shrieks and a third added shrieks, wails, hoots, and intergradations of all three. Often it seemed possible for me to count the number of vocalizing individuals in a chorus by ear if three or fewer. I could not reliably count larger parties by ear, but chimpanzees may well be able to estimate chorus sizes (see Bygott 1979:412).

Pant-hoots lasted from $\leqslant 5$ to approximately 15 sec; sometimes series of repeated pant-hoots were given which lasted several minutes. I estimated that some pant-hoots were audible for up to 2 km, and possibly even farther if the listener was on a hill. Audibility at 2 km advertises the presence of the caller over an area of 12.5 km^2. Of all the auditory signals given by forest animals at Ngogo, including elephant and hyena, the pant-hoots of chimpanzees were the most striking and seemed to carry the longest distance (see Wiley and Richard 1978 for an analysis of the acoustical physics of auditory signals in varying habitats).

Variation in style of pant-hooting between chimpanzees was apparent. I gradually learned to recognize those of a few individuals, but such recognition was often impossible to verify when the vocalizers were a great distance away. Recording the vocalizations of Gombe chimpanzees, Marler and Hobbet (1975) found significant differences between the pant-hoots of adult females, those of adult males, and those of juveniles. From an analysis of sound spectrographs they concluded, ''On this basis there are cues available in pant-hooting for individual recognition, although as with *Cercopithecus* sounds (Marler 1973), the likelihood that some properties of pant-hooting are more conspicuous than others to listening chimpanzee reduces the value of a simple statistical analysis of individual differences.''

Temporal distribution of pant-hooting

Daily occurrence. Figure 17 illustrates the collective distribution of pant-hoots on a diel basis. To collect these data I trained myself to awaken during the night when I heard pant-hooting and vocalizations of other animals and then I recorded the pertinent data. I was a light sleeper because for my first few months at Ngogo I slept in a mud hut with no door. Being asleep, though, is a sure way to miss behavioral events even if they are audible, so my record of the occurrence of pant-hoots between 2230–0600 hours is almost certainly underrepresented.

Two peaks in pant-hooting are prominent, one between 0600–1000 or

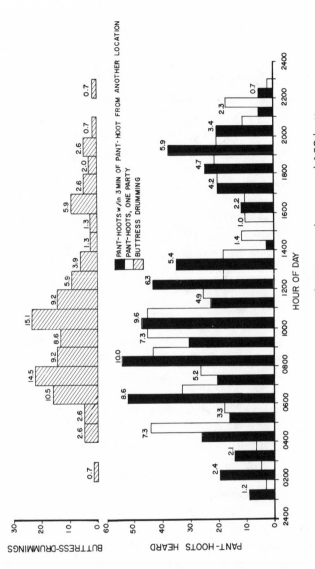

FIGURE 17. Hourly distribution of 962 pant-hoots and 152 buttress drummings heard at Ngogo between 21 April 1977–10 May 1978. Quantity at head of column indicates percent of total sample.

1300 hours and another around 1900 hours. Buttress drumming during pant-hooting was primarily a morning phenomenon. My impression was that pant-hooting, along with drumming, was most likely to occur during travel in the morning. Midafternoon usually was quiet; then, around nesting time, pant-hooting, but not drumming, again was common. This led me to suspect that much of the evening pant-hooting was done by apes in their nests. Some drumming did occur during darkness, but whether this was done by chimpanzees who climbed down from their nests merely to enhance an auditory display or was done by moonlight travelers is unknown. Goodall (1968) reported that a few individuals occasionally, but rarely visited the provisioning area at Gombe during the night, but she did not mention buttress drumming after nightfall.

Seasonal occurrence. Seasonal variation in the occurrence of pant-hooting between April 1977–May 1978 at Ngogo was marked (figure 18). Initially I suspected there might be a positive correlation between frequency of pant-hooting and fruiting periodicity, but an analysis of phenological data with the total frequencies of pant-hooting per month (excluding October 1977, which was not sampled adequately for vocalizations) indicated no significant correlation (correlation coefficient, $r = 0.2332$, d.f. $= 12, p > 0.05$).

The next logical correlation I suspected was between pant-hooting frequency and the presence of chimpanzees, as determined by numbers of individuals I encountered per hour in the forest. An analysis of monthly means again indicated no correlation (correlation coefficient, $r = 0.3673$, d.f. $= 12, p > 0.05$). A further comparison of frequency of isolated pant-hoots (i.e., those which were neither answers to a pant-hoot ≤ 3 min previously nor were answered by another pant-hoot within that amount of time) with mean monthly party size of chimpanzees also indicated no significant correlation (correlation coefficient, $r = 0.5229$, d.f. $= 12, p > 0.05$). Frequency of total pant-hoots heard, however, was significantly correlated with mean party size of chimpanzees seen per month (correlation coefficient, $r = 0.7107$, d.f. $= 12, p \leq 0.01$). An even higher statistical correlation exists between frequency of pant-hoots occurring within 3 min of pant-hoots from another location and mean party size seen per month (correlation coefficient, $r = 0.7851$, d.f. $= 12, p \leq 0.01$). The upshot of these analyses is that when chimpanzees traveled in larger parties they were more likely to pant-hoot, and even more likely

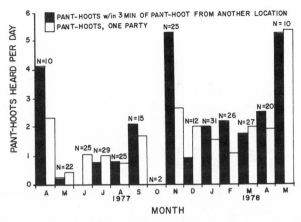

FIGURE 18. Monthly distribution of 965 pant-hoots heard at Ngogo between 21 April 1977–10 May 1978. Quantity at head of column is number of days of observer presence at Ngogo per month.

to pant-hoot in response to other pant-hooting. The seasonality of these trends (larger party sizes and more pant-hooting) I suspect was related to fruiting periodicity, but analysis of my phenological data do not indicate a significant correlation perhaps because those data were not comprehensive enough.

Context of long-range calling

Pant-hooting occurred in a variety of contexts, most of which seemed socially motivated: when approaching a food source, when traveling, while watching another party approach, in response to the pant-hooting of another party, and apparently while lying in a night nest. In several instances two or more of the above elements coincided with pant-hooting. Pant-hooting often seemed to be elicited by exciting situations, e.g., approach to food and calls or approaches by other chimpanzees, but often it apparently occurred spontaneously or unexpectedly from my perspective.

Pant-hooting was predominantly a male activity (table 34). I saw adult females initiate pant-hoots on only ten occasions, all of which were in the context of feeding in a large fruiting tree, nine of which also were in the context of watching another adult chimpanzee approaching, as if in

TABLE 34. NUMBER OF INCIDENTS OF PARTICIPATION IN PANT-HOOTING PER HOUR OF OBSERVATION BY CHIMPANZEES IN KIBALE FOREST, UGANDA.

Sex of vocalizer	Time observed 1976–1978 hour:min	Individual initiating pant-hoot vocalizations/hour	First pant-hooting responder to distant (>50 m) pant-hoot vocalizations/hour	Non-pant-hooting responder to local (<50 m) or distant (>50 m) pant-hoot
Male	289:10	0.211 (85.9)	0.041 (66.7)	0.387 (54.6)
Female	484:35	0.021 (14.1)	0.012 (33.3)	0.192 (45.4)
Total	773:45	0.092[a] (100.0)	0.023[a] (100.0)	0.265[a] (100.0)

NOTE: Only adults and subadult males included in analysis. (Quantity) is percent of total pant-hooting incidents. See text.

[a] Averages.

greeting. Adult females seemed to greet adult males with pant-hooting, but greeted other females with lower intensity pant-grunts; juveniles usually were not greeted.

Males were significantly more likely to initiate a pant-hooting sequence and to vocally respond to a distant pant-hoot than were females ($\chi^2 = 22.351$, d.f. $= 2, p \leq 0.0005$). Oddly, adult males were also more likely than females to be nonresponders to distant pant-hooting. Although it may seem paradoxical that males were more likely to be responders *and* nonresponders to distant pant-hooting than were females, the data make sense when one considers that while I observed adult males (37.4% of adult observation hours during the period of collection of pant-hooting data) often there were few or no females nearby, but distant pant-hooting was commonly heard (probably pant-hooting by other males?). And when I observed females, often there were no males evident and very little pant-hooting, nearby or distant, to which the females might respond. In other words, males had more opportunities to be responders and nonresponders to distant pant-hooting than did females and scored higher on both counts.

These data suggest that pant-hooting was a mode of communication between males who were separated from one another in the forest. If individual males recognize one another's vocalizations, which seems probable, then when they respond to distant pant-hooting they know to whom they are responding and, conversely, they may know the vocalizers to whom they do not respond. Special affiliative relationships between males, some of whom probably were siblings, have been reported by Goodall (1968), Bygott (1979), Nishida (1979), and in this study. Pant-hooting may facilitate reunions between these individuals. Further research is necessary to test the hypothesis that adult males selectively respond to the long-range calls of known individuals.

Contagious effect of long-range calling

Table 35 indicates that 32 percent of all pant-hoots were given by a single chimpanzee only. The remaining 68 percent of recorded incidents of pant-hooting were joint vocalizations of 2–$\geq$8 individuals. Rarely, if ever, did a chorus begin as a synchronized vocalization; one ape usually was the initiator. It was my impression that many chimpanzees joined into pant-hooting choruses that they never would have initiated. It was as

TABLE 35. ESTIMATED NUMBERS OF CHIMPANZEES GIVING PANT-HOOTING CHORUSES AT NGOGO, KIBALE FOREST, UGANDA, BETWEEN APRIL 1977–MAY 1978.

No. of chimpanzees in chorus	N	Percent of total	Incidents per hour of day											
			2400–0059	0100–0159	0200–0259	0300–0359	0400–0459	0500–0559	0600–0659	0700–0759	0800–0859	0900–0959	1000–1059	1100–1159
1	308	32.0		9	10	12	11	9	16	20	35	31	41	12
2	256	26.6		2	3	6	19	21	30	10	20	8	23	10
2–3	282	29.3		1	10	2	32	1	29	11	28	19	14	11
≥3	79	8.2					8	1	7	9	9	15	11	3
4	8	0.8									3			
5	14	1.5							1		3	1	1	5
6	9	0.9												6
7	2	0.2												
≥8	4	0.4											2	
Total	962	99.9	0	12	23	20	70	32	83	50	98	74	92	47

No. of chimpanzees in chorus	Incidents per hour of day											
	1200–1259	1300–1359	1400–1459	1500–1559	1600–1659	1700–1759	1800–1859	1900–1959	2000–2059	2100–2159	2200–2259	2300–2359
1	14	10	4	4	10	9	15	17	12	6	1	
2	4	11	5	4	9	3	22	24	10	6	6	
2–3	35	27	3	2	1	23	5	9	9	10		
≥3			1		1	5		7	2			
4	1	2					2					
5	2	1										
6	3											
7		1					1					
≥8	2											
Total	61	52	13	10	21	40	45	57	33	22	7	0

NOTE: Chorus sizes ≥4 are actual counts not estimates.

if being in the midst of a full-blown pant-hooting vocalization was too compelling to resist or ignore.

Not only were pant-hoots difficult to ignore for chimpanzees in the same aggregation as the initiator, they often were compelling to other chimpanzees some distance away from the vocalizing party. Figures 17 and 18 illustrate the tendency for pant-hoots to occur more often within 3 min of another pant-hoot emanating from another location (53.2% of all pant-hoots) than for pant-hoots to occur isolated in time from a single location (46.8%). It is difficult to compute the random probability for frequency of occurrence of pairs or greater multiples of pant-hoots from two or more locations as compared to the probability for isolated pant-hoots. The mean number of pant-hoots recorded for all days of recording ($N = 279$) was 3.46 calls/day. The probability that multiples of pant-hoots should occur within 3 min of one another at different locations must be very small compared to the random probability that pant-hoots should occur independently at intervals ≥ 3 min apart. Of course pant-hoots were not random events. They depended upon the nonrandom presence of chimpanzees and upon eliciting stimuli in specific contexts that are not understood completely. But it is clear that the majority of pant-hoots were not random and either elicited or were elicited by other pant-hoots.

Functions of long-range calling

Communication of identity and location. Pant-hooting by one or more chimpanzees broadcasts specific information about the number, possibly the identity, and the location of the vocalizing individuals over an area up to, and possibly greater than 12.5 km² at Ngogo. Most pant-hootings elicited others that conveyed similar information. This long-range calling suggests genuine long-range communication between individuals separated from one another in a habitat where visual signalling is limited to short distances. This applies especially to adult males. Bygott (1979:410) came to this same conclusion:

Sometimes displays appeared to be a response to the distant calls of other chimpanzees; here, however, 34% of vocal displays and only 16% of nonvocal displays began within 5 min after distant calls had been heard. This data [sic] further confirms the suggestion that vocal displays are mainly a nonaggressive form of long-range communication. Often, a lone male or a party of males would perform several vocal displays in succession, pausing to listen carefully after each display, apparently trying to evoke a response from other chimpanzees within hearing range.

The function of pant-hooting choruses may be similar to howling by wolves: to communicate the locations and identities of pack members, who often become widely separated from one another (Murie 1944; Joslin 1966; Mech 1970). In the fusion-fission social system of chimpanzees, individual associations must be of value to the individuals concerned, or they would not be so frequently or elaborately reformed (see Bauer 1979). Long-range calling probably facilitates the reunion of separated individuals in an environment where visibility rarely is greater than 100 m, and where olfactory signaling would be unreliable and too vague for individuals who travel extensively each day. In addition, long-range calling would inform alien chimpanzees of the presence of conspecifics and may function as a territorial marker.

Food calling. Wrangham (1975) reported food-calling by adult male chimpanzees at Gombe. Food calls were defined as a specific, often distinguishable type of pant-hooting given by an adult male as he approached a large food source. When food calls were given, new individuals were significantly more likely to arrive (33%) during the bout of the food caller than if he had not called (6%). Independently and unaware of Wrangham's report, I observed the same phenomenon at Ngogo.

When arriving at patches of vigil size at Ngogo, one or more adult male chimpanzees of a party sometimes (23.7% of $N = 93$ visits) vocalized with long-range calls. I defined a food patch as one of vigil size if it was a tree containing a superabundance of fruit and was visited repeatedly by several primate species and other frugivores (see discussion of patch size in chapter 4). "Superabundance" implies nothing about the presence or absence of competition among foragers; it means only that ripe fruit was extremely abundant. The long-range vocalizations given by Ngogo males were pant-hoots characterized by wide variance in sounds both between and among vocalizers. Rhythmic hoots, grunts, shrieks, screams, wails, and an occasional roar were juxtaposed, and often the vocalizer pounded on the surface of a tree buttress just before climbing up to feed. On one occasion an adult male, Raw Patch, kept up almost continuous pant-hooting for 12 min, beginning 2 min before he climbed into a large, fruiting *F. dawei*. The two males who followed him were silent. I never saw a female initiate such calls in a similar context.

While I was never able to interpret distant pant-hooting as food calling with certainty, H. Plooij (cited by Wrangham 1975) reported that the "food pant-hoot" at Gombe was distinguishable from other types of pant-hoots.

Perhaps the initial question arising in this context is whether or not alleged food calls attract other chimpanzees who subsequently share the food resource of the caller. To test this question I pooled all my data pertaining to arrivals of parties containing one or more males at vigil trees in fruit, excluding cases when the party deserted the tree prematurely in response to my presence. Table 36 summarizes data on arrivals of parties of chimpanzees at vigil trees following the primary arrivals of parties containing one or more adult males who either gave food calls or did not. Arrival of secondary parties was significantly associated with food calls ($\chi^2 = 4.049$, d.f. $= 1$, $p \leq 0.05$). Food calls evidently attracted other chimpanzees, both male and female, who subsequently shared the food resource of the caller, thus confirming Wrangham's (1975) earlier report.

It is tempting to conclude from observations of interindividual distances in patches (chapter 4), and from calling by adult males and responses to calling at food patches, that chimpanzees are highly cooperative, perhaps even altruistic in their foraging. Their calling behavior seems to increase the callers' competition for a known food source, and the changes in interindividual distances with an increasing number of foragers suggest that competition does occur within food patches. The behavior of chimpanzees seems to minimize aggressive interactions, and there-

TABLE 36. FREQUENCY OF ARRIVAL OF ADDITIONAL PARTIES OF CHIMPANZEES AT SUPER-ABUNDANT FOOD SOURCES FOLLOWING "FOOD CALLS" AND NO "FOOD CALLS" BY ADULT MALES AT FOOD SOURCE, NGOGO, KIBALE FOREST, UGANDA, BETWEEN JANUARY 1977–MAY 1978.

		Arrival of subsequent chimpanzee party during feeding bout of caller, or within 30 min of call.		
		Yes	*No*	*Total*
Arriving party containing ≥ 1 adult male chimpanzee who:	Pant-hooted upon arrival	13 (59%)	9 (41%)	22 (100%)
	Did not pant-hoot	18 (34%)	35 (66%)	53 (100%)

fore lowers the interference between apes while feeding. Even so, a chimpanzee acting in its own interest should not attract others to a food source in which he may need to reduce his own consumption as a result, despite the competitive edge adult males have because of their dominance status. If there is competition among individuals foraging in a patch, calling behavior takes on the aspect of altruism.

There are alternate explanations of the behavior, however. Pant-hoots probably were identifiable individually (Marler and Hobbet 1975). Males may have been pant-hooting to mark their territory. Conversely, individuals who appeared in proximity to a pant-hooting adult male shortly after his vocalization may have been attracted to the male himself rather than his food resource. But even if this were the case, the result is that they consume resources which perhaps would have been consumed by the caller in the absence of later arrivers.

If this competition were significantly deleterious to the callers, selection should extinguish calling behavior at superabundant food patches. But calling occurs somewhat predictably there. In fact, Wrangham (1975) found that Gombe males were more likely to call when arriving at quantitatively large than at small food resources. Probably an adult male is capable of assessing the amount of food before calling. Unless he overestimates that amount, the cost of competition is probably trivial. In short, my observations indicate little to suggest that callers invite serious competition.

The benefits to having called derive from three potential consequences: (1) the caller may benefit by social interaction with a subsequent arriver—e.g., by mutual grooming or being in a larger party for safer patrolling of a territorial boundary; (2) the subsequent arriver may be an estrous female with whom the caller may mate; or (3) the subsequent arriver may be related to the caller, which is likely if it is another male or an immature female (assuming a female exogamous system), so that any increment added to the individual fitness of the related arriver by improved nutrition is shared in the inclusive fitness of the caller.

In conclusion, the phenomenon of food calling seems to be an adaptation of males which, because of the component of genetic relatedness necessary for some of its advantages, probably arose through kin selection among males.

7

THE SOCIAL ECOLOGY OF CHIMPANZEES

The ecology of Kibale chimpanzees consists of the exploitation of irregular patches of rare and ephemeral resources by specialist frugivores in a competitive environment. Considering the biomass of competing species, competition was primarily interspecific. The distribution and levels of availability of the apes' food resources is in constant seasonal and aseasonal flux. The areas in which a chimpanzee ranges during a day or a lifetime are in part determined by the spatial and temporal distribution of these critical resources, but ranging is also influenced by the dispersion of conspecifics with whom the individual must socialize to maximize its individual and inclusive fitness.

My interpretation of the social structure of Kibale chimpanzees as a product of ecological pressures and sexual and kin selection draws on data not only from Kibale, but also from the longitudinal studies of provisioned chimpanzees in Tanzania (e.g., Goodall 1971; Nishida 1979) and of wild chimpanzees elsewhere. This interpretation also addresses the predictions of the model of sexual selection and territoriality outlined in the introduction.

Associative patterns among the chimpanzees of Kibale Forest, Gombe, Mahale Mountains, and other regions indicate clearly that social groupings are temporary. The ephemeral aspect of chimpanzee sociality misled early workers who tried to make sociological sense from their field observations (Nissen 1931; Azuma and Toyoshima 1962; Kortlandt 1962; Goodall 1963; Reynolds 1963; Sugiyama 1968). The travel, coalescence, and division of parties, and exchange of members between parties—often accompanied by vigorous displays and a cacophony of piercing vocalizations—were an anathema to those who anticipated a well-defined, discrete social structure among chimpanzees similar to other primate social systems known at the time. The lack of such a strict social order led Goodall (1963), Reynolds (1963), and Sugiyama (1968) to postulate that chimpanzee society was open and panmictic over the species' contiguous range, a kind of species-community characterized by local populations, the individuals of which were more likely to socialize together because it was spatially more convenient to do so. That the apparently loose social system of chimpanzees was a complex adaptation shaped by the pressures of an unusual ecological niche, that of a large-bodied, social frugivore, and by sexual selection was not clear.

Memberships of social groupings of chimpanzees are temporary. An individual's decision to remain with or split from a party probably is made by simple cost–benefit rules weighing the benefits of socializing against the current availability of food and the cost of competition by companions. A large traveling party incurs high competition, even for individuals of high status, because large food trees are rarely defended against companions. Unless resources are plentiful, socialization may be too expensive. As food becomes more sparse or dispersed, but party size remains constant, the daily range of each of its members must increase to maintain the nutritional modus vivendi. The alternative to increasing the daily range of the party to visit more patches is to split from the party, disperse, and travel in a smaller party or alone. The lower nutritional demand per social unit requires that fewer patches be visited.

Chimpanzee society, as characterized by fissioning and fusing, is an individual adaptation to maximize foraging efficiency. Individuals who can split away from a social group to forage on scarce resources during lean times and *still* maintain a complex social network have an advantage over those who favor large parties when available resources are inadequate. Fusion-fission sociality allows these large frugivores to exploit dispersed patches of fruit which, for the community to exploit as a unit, would be prohibitively expensive bioenergetically. Individuals who travel and forage together do so only when dispersion of food allows it. The degree to which individuals can prolong party cohesiveness by taking advantage of the superior foraging experience of others, as Horn (1968) has suggested for colonially nesting birds, is not known.

Fusion-fission sociality remains one of the most striking aspects of chimpanzee ecology. As evidenced by the large aggregations gathered in the provisioning area at Gombe before 1969 (Goodall 1968; Wrangham 1974; Reynolds 1975) and gathered around other superabundant food sources during other studies (e.g., Reynolds and Reynolds 1965; Nishida 1968; and this study) chimpanzees join into larger-than-average traveling parties and socialize more with one another when normal feeding constraints are lifted. Despite all bioenergetic constraints imposed by the distribution of food, the chimpanzees of a community maintain a social network that persists over lean seasons.

This phenomenon asks some fundamental questions: (1) To what extent is chimpanzee social structure a predetermined product of ecology and of phylogeny? (2) What is the precise nature of the social network, or community of chimpanzees? and (3) Why is the community important, or, in other words, what do individual chimpanzees gain by socializing within the larger structure of the community?

EVOLUTIONARY PRESSURES SHAPING CHIMPANZEE SOCIAL STRUCTURE

The social structure of a population and its ecological relationship to its habitat are intertwined products of evolutionary forces acting over time. Several schematic attempts have been made to correlate social systems of primates with the habitat type in which they occur. In a pioneer effort Crook and Gartlan (1966) listed five grades of increasing social structure and group size in correlation with a transition of habitats from rain forest to wooded or arid savanna, concomitant with a gradient of increasing variability and tolerance with regard to the type and condition of food types used. Grade I, for example, primarily contained sexually monomorphic, nocturnal, solitary, insectivorous, forest-dwelling prosimians, while Grade V was represented by sexually dimorphic, diurnal, large-group living, omnivorous, arid savanna dwellers such as *Papio hamadryas* and *Theropithecus gelada*. But this scheme, based on gross habitat types, was riddled with exceptions and lacked explanatory power.

One of its problems was that habitat types were based solely on vegetative physiognomy, ignoring structure, diversity, dispersion, and seasonality of food, successional state of vegetation, predation, and competition. The usefulness of habitat type to predict primate social structure is illustrated by Rodman's (1973b) observations of five sympatric species of Bornean primates. These exhibited social systems extending through Crook and Gartlan's (1966) Grades I to IV, yet all lived within 3 km^2 of dipterocarp rain forest. Rodman (1973b) asked appropriately, "Why should we expect to find a modal 'rainforest' type of social organization if our basic postulate is that social organization is related to the habitat and the rainforest is a *different* habitat for each species found in it?" (italics mine). To further refine the classification of habitats into ever more distinct subhabitats, somewhat along the lines of the "*n*-dimensional hypervolume" niche of Hutchinson (1957) makes a habitat–social grade scheme unwieldy.

Working within a single habitat type Struhsaker (1969) observed rain forest species of African Cercopithecinae to test the extent to which social organization was a function of ecology versus phylogeny. Struhsaker (p. 113) concluded:

> In considering the relation between ecology and society it must be emphasized that each species brings a different phylogenetic heritage into a particular scene. Consequently, one must consider not only ecology but also phylogeny in attempting to understand the evolution of primate social organization. The interrelations of these two classes of variables determines the expression of the character, in this case social structure.

In other words, phylogenetic inertia may prevent some species of primates from achieving the maximum conceivable population densities they might attain in certain habitats *if* they were to alter their basic social structures. A more exhaustive treatment along the same lines, but in Kibale Forest (Struhsaker and Leland 1979) reaffirms this conclusion.

Eisenberg et al. (1972) synthesized another social-grading system correlating primate social structure (characterized by numbers of males in a social group: unimale, age-graded male, or multimale classes) with feeding ecology, e.g., insectivore-frugivore, arboreal folivore, semi-terrestrial frugivore-omnivore, etc. Considering the worldwide radiation of primates, they also concluded that phylogenetic inertia was a stumbling block to neat correlations based on ecology alone (p. 873), ''no doubt some of the differences in social structure seen in those species adapted to similar ecological niches will be found to result from phylogenetic differences; that is, the social structures of the ancestral forms have been carried forward in the adaptive radiation of these species.'' The classificatory scheme of Eisenberg et al. also was riddled with exceptions and weak in predictive power. Clutton-Brock (1974:539) further elucidated the reason why the classificatory schemes outlined above and elsewhere (Crook 1970; Kummer 1971) lacked predictive power:

> Different species tend to react to similar environmental pressures in different ways. When a novel adaptation evolves, its form will be partly determined by the various environmental factors through which selection is operating and partly by phylogenetic inheritance. Consequently, distantly related species are likely to evolve different traits with similar functions. . . . many interspecific differences in social organization may well prove to represent different methods of overcoming the same ecological problems.

Phylogenetic inertia, however, has not prevented the evolution of three divergent social systems among the three species of great apes. The basic social unit of mountain gorillas, which is similar to that of lowland gorillas (Jones and Sabater Pi 1971), is essentially a one-male harem, although occasional groups contain more than one fully adult male for periods of time (Schaller 1963; Harcourt 1979a,b; Fossey 1981, 1982; personal observation). But the habitat of the folivorous mountain gorillas is montane forest containing plant communities (such as bamboo forest) which are floristically simple relative to the habitat of chimpanzees and are apparently unsuited to the frugivores. Mountain gorillas provide evidence that phylogenetic inertia did not prevent evolutionary divergence in social structure from the ancestral species of modern great apes (an ancestor unknown to archeologists of today). But because of the difference in habitat types, the ecological determinants of social structure among the great apes remain unclear. For a valid ecological comparison with chimpanzees, orangutans provide the next phylogenetically close species (Miller 1977).

Ecologically chimpanzees and orangutans are very similar (MacKinnon 1974; Galdikas 1979; Rodman 1979). Orangutans are highly frugivorous great apes of Sumatran and Bornean rain forests. Patterns of food availability seem similar for both species, and competition with sympatric frugivores may be equally severe. In contrast to the highly social chimpanzees, orangutans are strictly solitary except during brief periods of mating (Rodman 1973a; MacKinnon 1974). Rodman (1973a, 1979) discussed an ecological sexual dimorphism that may be responsible for males and females foraging separately. The body size of an adult male is approximately twice that of an adult female. Males have greater metabolic needs (see Gaulin 1979) which preclude their foraging as selectively as females, who travel further each day and tend to feed less at each patch. Yet in other ways orangutans and chimpanzees seem to face similar ecological problems. The primary difference between these two species is mobility, particularly the ease and efficiency of terrestrial locomotion (see Rodman in press for an independent discussion of this). As a behavioral indicator of mobility, day ranges, considering only solitary individuals, are much longer among chimpanzees (c.f. Rodman 1973a and MacKinnon 1974 with Wrangham 1975). The superior mobility of chimpanzees ameliorates constraints of foraging enough for individuals to associate on a daily basis during favorable seasons.

Comparisons with gorillas and orangutans illustrate that neither phylogeny nor ecology per se can account for the social system of chimpanzees. Goss-Custard et al. (1972:1) focused on the theoretical rationale necessary to understand the primary pressures shaping social systems. Noting that previous authors considered the fundamental features of social organization "largely in terms of the *survival* of individuals or groups," they elected instead to "concentrate more on the bearing that variations in social structure have on the reproductive success of individuals." They examined, therefore, "the possible effects not only of ecological selection pressures but also of sexual selection."

Evolution occurs because of the differential reproductive success of different genotypes over generations. Primate social systems cannot be understood as a product of evolution without considering the processes of evolution, particularly sexual selection. Because reproduction in primates involves a much greater investment of time and energy by females, males find themselves with a potential "surplus" of parental investment (Trivers 1972) per offspring vis à vis the females with whom they mate. For this discussion parental investment should be considered as that portion of each parent's life span and that amount of biological energy necessary to reasonably ensure that an offspring reaches reproductive maturity. Under favorable conditions, in which a hypothetical female primate can successfully raise her offspring with little or no assistance from a male, the parental investment of males may amount only to the time and energy spent in courtship and insemination. In mating systems in which females give greater parental investment per offspring than males the females become a limited resource for the males because males are not limited to equally matching parental investment per offspring with a single female. In systems where females invest more it is in a male's greatest reproductive interests to inseminate as many females as possible, and, other traits being equal, natural selection will favor male genotypes which excel at this. Because the total reproductive success of one sex is equal to that of the opposite sex during each generation, the insemination of multiple females by a single male—to achieve higher than average reproductive success among males—will result in competition between males over opportunities to mate with reproductive females. This competition leads to sexual selection.

An increase in parental investment by an individual of one sex may mean a concomitant decrease in the amount of parental investment necessary by its mate to produce viable offspring with chances of survival equal to the old balance of parental investment. But it also could mean a consequent higher total reproductive success for the parent of the second sex should it *not* reduce its parental investment. Therefore, *each* sex should try to induce the second to maximize parental investment. This relationship may not hold true if increased parental investment by the parent of the second sex lowers the reproductive success of the parent of the first sex (and, hence, of both sexes). For example, if a male remains with a female after he has mated with her, she will have to compete with him for food, and such competition may be worse for her reproductively than if she were solitary. This may be the case among orangutans (Rodman 1973a).

In many species of mammals, however, parental investment by males is considerable and apparently enhances the reproductive success of both sexes: for example, *Callicebus* monkeys (Mason 1968), white-handed gibbons (Ellefson 1968), wild dogs (Kuhme 1965; Estes and Goddard 1967), wolves (Murie 1944; Mech 1970), and Hartmann zebra (*Equus zebra hartmannae*) (Joubert 1972) to name only a few. In general, natural selection will favor female phenotypes which, by whatever means, induce males to contribute the maximum amount of parental investment to their offspring which enhances that female's reproductive success. The form and amount of advantageous parental investment by males depends on the ecological situation faced by the species. Among Kibale chimpanzees ecological constraints somewhat similar to those faced by orangutans, but ameliorated by the superior mobility of chimpanzees, dictated a major parental investment by females.

The concepts of parental investment and sexual selection are a necessary framework for describing the basic social network of chimpanzees (see also Wrangham 1979b).

THE SOCIAL NETWORK, OR, CHIMPANZEE COMMUNITY

After two decades of field research on chimpanzees the nature of their basic social system is still unresolved. Because the basic social network or community of chimpanzees apparently almost never is together in a single aggregation and because parties of chimpanzees change in membership frequently, often with little or no ritual exchange between party members, it is not possible for a human observer to generalize after diligently following a social group and counting heads. The superficial randomness of associations between individuals is misleading. Kibale chimpanzees associated with one another nonrandomly with respect to age-sex-reproductive classes and to individual identities. Their social order resulted from the tendency for like-sexed individuals to reform associations frequently while other types of associations occurred less frequently or not at all.

Wrangham (1979a) analyzed the associational tendencies and traveling patterns of Gombe champanzees and tried those data for closest fit to three hypothetical models of chimpanzee community structure. The first, "classic primate" model assumed uniform ranging by all age-sex-reproductive classes within a well-defined and somewhat equally visited home range. Because adult females ranged less distance than adult males or estrous females, and they used smaller and more separate core areas than did the latter age-sex classes, Wrangham concluded that the first model did not fit the data.

In the second model all individuals occupied a single, large home range over which adults and possibly estrous females roamed more or less completely, but in which adult females with infants maintained much smaller, overlapping core areas epicentered in the core of the large home range. But because males migrated or otherwise shifted their range independently of females with infants (see also Kano 1971; Nishida and Kawanaka 1972; Nishida 1979), Wrangham (1979a) concluded that the second model was not accurate. The home range was not the same piece of real estate for both sexes.

Wrangham proposed a third model in which adult females with infants were dispersed more or less uniformly over suitable habitat and independently of the patterns of ranging of adult males, who together traveled over a much greater home range than did any of the females with dependent offspring. In this model males ranged within an area with roughly defined boundaries which separated their home range from those of neighboring communities of males. The females in this model ranged as if the male home range held no significance for them. The core areas of some females could straddle the boundary between home ranges of neighboring groups of males. Because actual observations of the shifting of the ranges of Gombe males did not include a concomitant shifting of the ranges of adult females with infants, Wrangham suggested that the third model accurately described chimpanzee social structure. Wrangham (1979a:489) concluded, "Sex differences in ranging behavior clearly exist; however, the current data favor applying the community concept only to males."

Wrangham's third model of chimpanzee social structure implies separate social systems for males, estrous females, and females with dependent offspring. The latter usually range independently of other matrilineal units and usually completely independently of adult males, unless they come into estrus, travel with a grown son, etc. A female's home range (core area) may overlap one or more home ranges of communities of males strictly at random (sociologically) in Wrangham's third model. This picture closely resembles the model of dispersion of female orangutans (Rodman 1973a; Galdikas 1979), a solitary species in which males apparently range over the home ranges of several females. Competition between male orangutans is intense at times, and it appears to be in a male's best reproductive interests to have access to (or be accessible to) as many females as is ecologically possible. The main difference between the social system of orangutans and Wrangham's (1979a,b) model for chimpanzees is that male chimpanzees maintain a *communal* home range overlapping the ranges of several females.

Although compelling, Wrangham's proposed model of chimpanzee social structure does not account for the degree of affinity between adult females that I observed at Ngogo. Females often traveled, foraged, and groomed together in contrast to their weaker associational tendencies with adult males. Adult female chimpanzees were *not* predominantly solitary, as were adult female orangutans (Galdikas 1979; MacKinnon 1979; Rodman 1979), but frequently ranged together, almost as if they were members of a female community. In view of this, the males-only community model of Wrangham (1979a) seems oversimplified.

Clearly adult females who leave their natal home range temporarily or permanently to breed with the males of another community (Nishida 1979; Pusey 1979) are behaving as though they had some concept of a male community; breeding with unrelated males is in their best reproductive interests (Wilson 1978:37–38). Observations of resident females chasing nonresident females (strangers?) away from the core area of their community home range (Goodall 1971b, 1977; Bygott 1979; Pusey 1979) indicates an "us-and-them" concept, and also indicates that amicable females of a communal home range forcibly exclude strangers with whom they otherwise would have to compete. Such communal action is inconsistent with the males-only model of Wrangham (1979a). Another factor to consider is Bygott's (1972) observations of adult male Gombe chimpanzees attacking unfamiliar females and killing their infants. Although such infanticide is adaptive reproductively (and, hence, explicable) from the points of view of the males because the females, if they remained, would come into estrus years earlier (Hrdy 1977, and 1979 for a general review; Fossey 1982; vom Sall and Howard 1981), considering the females who were attacked and whose infants were killed, the concept of community as strictly a male phenomenon seems biologically inaccurate. It would be in a female's best reproductive interests not to frequent a boundary area between communities if she has dependent offspring because of the risk of attack by alien males. In short, each female should "identify with" or be a member of a specific community. Hence, the community cannot be considered realistically as strictly a male phenomenon. Females do not range along with males because of their different foraging priorities and reproductive strategies (see chapters 4 and 6, and the section below).

Unfortunately my data on ranging from Kibale are too few to analyze relative sizes, degree of overlap, and synchrony in utilization of ranges of males and females to further test Wrangham's (1979a) community model. Other data, however, shed light on it. Based on measures of affinity and socialization (e.g., traveling companions and grooming partners) between individuals, three points stand out: (1) males exhibited the highest degree of affinity toward one another; (2) females exhibited affinity toward other females, more so than toward males; and (3) males associated with females in ways not related to immediate interest in mating. This socialization and affinity is difficult to interpret within the framework of Wrangham's model, which describes females as being somewhat evenly dispersed throughout the habitat, rather than as social individuals in a fluid fusion-fission society who range solitarily when ecological conditions dictate but feed and sometimes travel together when food is clumped in large patches.

Data from this study suggest that both males and females are members of the basic community, or perhaps are members of two unisexual communities in the same home range. The principle of parsimony, however, suggests that the resident females and males of the males' home range are members of the same community, which probably is the natal community of the males and some of the females (Pusey 1979), rather than of two communities occupying the same space. To summarize, Kibale chimpanzees apparently socialized within a disexual community structure. Reports from Mahale Mountains (Nishida and Kawanaka 1972; Nishida 1979), Budongo Forest (Reynolds and Reynolds 1965; Sugiyama 1968, 1969), and Gombe National Park (Goodall 1971a; Goodall et al. 1979; and several other sources) indicate a similar picture for their respective communities.

BENEFITS ACCRUING TO THE SOCIAL CHIMPANZEE

The final question posed by chimpanzee fusion-fission sociality is how the community structure benefits its individual members. In contrast to many species of primates (Jolly 1972), chimpanzees neither form nor seem to benefit from a stable large group. Large-group size, a primary counterpredation strategy (Hamilton 1971; Busse 1977), apparently is of little advantage in the classical sense because chimpanzees face few predators other than human hunters. Against humans large group size probably would be more of a liability than an advantage because it would increase the detectability of the whole group. Thus, there is little benefit for a community to remain spatially coherent, and there is a large bioenergetic cost because of the increased travel that is necessary to maintain such coherence. And female chimpanzees, like female orangutans, would benefit little if the adult males with whom they mated were to remain with them to protect them against predators.

In Kibale Forest the parental investment by male chimpanzees apparently consisted primarily of insemination of a female, but it did not end there. Juvenile males often traveled with adult males, who may have been their older siblings or fathers in some cases. If the society was male-retentive, these males certainly were more closely related to one another than to the average adult female. In this way males may have been increasing their parental investment toward male offspring. Juvenile females of the same age apparently remained with their mothers. Additionally, adult males may have increased their parental investment by maintaining a community territory excluding competition by alien chimpanzees.

I observed no overt territorial clashes in Kibale, though. Perhaps this was because the study areas I used were not in regions of overlap between adjacent communities. Perhaps instead, Kibale chimpanzees were not territorial. But males behaved as though the model of sexual and kin selection and territoriality outlined in the introduction was accurate. The predictions it generated were verified by data from Kibale:

(1) Adult males traveled more frequently with other males, both mature and immature, than they did with females. One explanation is that males were traveling together as a means of increasing the adults' reproductive success by donating more parental investment toward immature offspring—by improving their educations in ecology, so to speak. But often only mature individuals traveled together. The most compelling explanation is that males benefited more from one another's company than from that of females because dominance in encounters between communities generally falls to the larger and more aggressive party (Nishida and

Kawanaka 1972). This is not to imply that all multimale parties were aggressive. Large party size is also simply safer in an environment of social aggression.

(2) Males apportioned affiliative behaviors (e.g., mutual grooming) more frequently and for greater duration toward one another than toward females. This could be the consequence of a kin-selected, male-retentive society and be explained again by the increased inclusive fitness which accompanies the apportionment of beneficial social behaviors toward one's kin. But social grooming, while hygienic, also tends to function in the re-forming and strengthening of social bonds (Simpson 1973; Bauer 1979). For males, who may communally and cooperatively patrol and defend a home range against alien males who are willing to fight, strong male-male bonds seem a necessity. Other studies also have indicated strong affinities between males (Simpson 1973; Bauer 1979; Nishida 1979; Pusey 1979; Wrangham 1979).

(3) Male chimpanzees traveled during a greater percentage of their active daily budget than did females. They also spent more time feeding. For metabolic reasons, females, especially females with dependent offspring, should travel the minimum distance for a daily itinerary that provides adequate nutrition for her and her offspring. Males in a territorial system need to go beyond the needs of daily nutrition to mark their home range by their physical presence and by giving pant-hoots, building nests, defecating, or otherwise leaving signs of their proprietorship of the territory. Males must travel beyond the demands of nutrition also as part of their reproductive strategy. Because estrous females are so rare, wide areas must be covered to increase the probability that males will encounter them. Theoretically, increased travel improves each male's chances of mating. But whether the increased travel by males was due to this aspect of reproductive strategy or to territoriality (in itself a more important component of male reproductive strategy) or both is unclear. In either or both cases such ranging by males is incompatible with the metabolic situation faced by adult females accompanied by immature offspring whom they must transport and feed.

In summary, relations between males in Kibale suggest but do not prove that territoriality was likely and that males affiliated disproportionately with one another as part of their reproductive strategy.

As noted above, from feeding ecology alone one probably would expect chimpanzees to exhibit a diffuse social system similar to that of orangutans, but the greater mobility of chimpanzees is associated with a very different pattern of reproduction and gene dispersal. Orangutans of both sexes usually disperse from their natal home ranges (MacKinnon 1974; Galdikas 1979). Benefits accruing to socializing adults of the same sex are not likely to include increased inclusive fitness. Among chimpanzees, though, apparently only females emigrate from their natal community; males remain (Nishida and Kawanaka 1972; Nishida 1979; Pusey 1979). In a system where males remain in a home range to mate and beget more males, who do the same for generations, eventually all males in the communal home range will share more genes in common than with males of other communities (e.g., Glass 1953). Cooperative and altruistic-appearing behaviors within a species are expected more frequently among such related individuals than among those sharing fewer genes in common (Hamilton 1964, 1972). Observations of two populations of provisioned chimpanzees indicate this high degree of affiliation between males extended to cooperation in communal defense of their home range from parties of alien males and other age-sex-reproductive classes (Bygott 1972; Nishida and Kawanaka 1972; Nishida 1979; Goodall et al. 1979). Those male chimpanzees seem a patent example of a kin-selected territorial breeding group. This system has its analogue in human societies. Morgan (1979) reported a significant trend among Yupik Eskimos to form whaling crews of umiaks on a kin basis. The hunts are dangerous. The Eskimos' explanation for why they want kin as crew members is because they can rely on relatives but not on nonclansmen in dangerous situations.

Brown (1969) outlined a general model to explain the evolution of territorial behavior:

> Its essence is that for territorial behavior to evolve in respect to a given object, be it mating priority, living space, foraging area or nesting site, 1) a situation must exist in which there can be aggressive competition for that object, 2) territorial individuals must be more successful than nonterritorial individuals in acquiring that object, and 3) the successful acquisition of the object of territorial behavior must raise the overall fitness of successful individuals over that of unsuccessful ones.

Note that Brown's model invokes Darwinian, or individual selection as a shaping mechanism for territorial systems because it possesses greater explanatory power than the group-selection model proposed by Wynne-

Edwards (1962, 1971) based on true altruistic cooperation, which lacks a demonstrable mechanism (Williams 1966; Wilson 1973). Some of the documented advantages accrued by territorial individuals include: buffering from disease and predation (Carrick 1963; Cody 1966), increased mating success (Peterson and Bartholomew 1967; Buechner 1971), a safe area for rearing young (Tinbergen 1956; Kuhme 1965), a reliable foraging area (Dice 1952; McNab 1963; Schoener 1968), and increased reproductive success (Carrick 1963; Schaller 1972; Bertram 1973).

Bands of male chimpanzees who patrol the boundary areas of their home range to display toward or attack alien males apparently maintain exclusive proprietorship of many square kilometers of home range vis à vis the males of other communities. Their territory contains all necessary resources for self-maintenance and reproduction, as in the "type D" sense of Hinde (1956) and to varying degrees may provide all of the benefits listed in the preceeding paragraph. Relations between these males are characterized by extensive mobility, high affiliation, cooperation on patrol, enhanced ability to exploit superabundant food sources due to food calling, and lowered competition for mating opportunities.

Lowered competition for mating opportunities is not surprising because of the high relatedness between males. An example of the strength of close kinship in facilitating cooperation between chimpanzees is provided by Riss and Goodall (1977). A partially paralyzed male, Faben, assisted his younger mature brother, Figan, to the position of alpha-male of the Kaskela community of Gombe by maintaining frequent proximity and joining in concert with Figan's aggressive displays during the latter's usurpation of the reigning alpha-male. Apparently the usurper, who was only of average size, might not have been successful had his brother not assisted. Tutin (1975) reported that the alpha-male at Gombe has a significant reproductive advantage, and Nishida (1979) reports similar findings from Mahale Mountains. In the reproductive strategy of chimpanzees nonsexual behaviors can effect reproductive success. By assisting his brother, Faben increased his inclusive fitness. It is possible that chimpanzees exercise a degree of choice using native judgement resembling Cody's (1966) *principle of allocation* to expend energy in social versus maintenance behaviors based on the ultimate reproductive consequences of the options. This does not assume that chimpanzees have foresight or intuition concerning the reproductive consequences of nonreproductive behaviors; this possibility rests only on the phenomenon of choice-making by chimpanzees about behaviors that ultimately *do* affect reproductive success.

On a wider scale, males who communally defend their home ranges and who are successful in expanding it at the cost of territories of other communities gain several reproductive advantages. They ensure a foraging area for the community's females and their offspring, they have exclusive access to breeding with those females, and perhaps gain access to new females. More females in the community increase one of the most limiting resources of the males' reproductive potential. Because the males are related to one another, each male shares in the increased reproductive success through increased inclusive fitness despite inequities in individual mating success. Natural selection should favor the kinship mating system of male chimpanzees provided the average reproductive success per male in the community exceeds that of males in a solitary system (see Gadgil 1972).

The benefits of territoriality to a female are less direct than those to males. Over time a female should be able to gauge the success of a community's males. If they are successful, she is assured of quality mates and a safe home range with adequate resources to raise offspring. If they are not successful males, she can decide to emigrate in search of a better community. Within the community home range where she eventually settles a female may exhibit a degree of territorial behavior toward unsettled females (Goodall 1971b, 1977; Bygott 1979; Pusey 1979), sometimes with the cooperation of other settled females. In this case females gain many of the advantages of territoriality that males do. Here again cooperation between adults may be vital. My observations of affiliations between females in Kibale make sense in this light.

The degree of social affiliation between Kibale females suggest the possibility of some form of kin selection among them. Such would not necessarily depend upon female fidelity to natal home range. If the females growing up in a community were to serially migrate to the same one or two neighboring communities to breed, then, over generations, the degree of relatedness between females of a shared home range would increase, possibly to the point where affiliative behaviors would increase. Current published data are too incomplete to test this idea. An alternative explanation for affiliative behaviors between females, other than reciprocal altruism, is that females are more compatible with one another based on activity patterns. It may be that a combination of these processes contributes to affiliative behaviors between females. Continued research might clarify this question.

SOCIAL AND ECOLOGICAL CORRELATES OF FUSION–FISSION SOCIALITY

Fusion-fission sociality is a rare system among mammals (Kummer 1971; Wilson 1975:137). Table 37 illustrates the diversity that do exhibit it. Doubtless other species will be added to the list as field work continues. Fusion-fission is defined here to describe social groups in which all members are rarely, if ever, together as a spatially discrete unit, and in which stable subgroups of specific adults do not recur daily, as one-male harems do. Consequently, hamadryas and gelada baboons (Kummer 1971) are not considered in this discussion. This section seeks to place chimpanzees in perspective with other fusion-fission societies.

Species exhibiting this form of social system have little in common. Phylogeny seems to play no part in its occurrence. Two ecological correlates universal with fusion-fission species are communal territoriality and patchy distribution of resources. The advantages of territoriality are discussed above; they vary between species. In conjunction with territoriality, though, patchiness may be a key determinant.

Patchiness is a relative and somewhat subjective term. What is patchy to a rodent may seem uniform to an elephant. The extent and significance of patchiness among the chimpanzees of Kibale were discussed in chapter 4 and above. But in general, when individual patches contain insufficient food for every member of a social group, more than one patch must be visited. If the bioenergetic expense of traveling to the required number of patches to feed all group members is prohibitive, the group must fission. If large group size remains an asset, for territorial defense, etc., the subgroups should fuse periodically to renew and strengthen social affiliations and to execute the large-group behaviors that make large group size an asset.

For spider monkeys (Cant 1978b) and chimpanzees the patches are fruit. For bottlenose dolphins (*Tursiops truncatus*) the patches are schools of pelagic fish (Würsig 1979). For wild dogs, gray wolves, coyotes (*Canis*

TABLE 37. SOCIOECOLOGICAL CORRELATES OF FUSION-FISSION SOCIALITY AMONG CHIMPANZEES AND SOME CLOSELY AND DISTANTLY RELATED SPECIES.

Species	Body size	Dispersion of food	Fusion-fission society	Territorial	Exogamy	All-male parties	Most common companions of females with infants	Source
Chimpanzee	large	patchy	yes	yes, males	female	common	females	Previous and present studies: see text.
Mountain gorilla	large	"uniform"	no, one male harem	no	both sexes	no	one male harem	Fossey 1979; Harcourt 1978, 1979a,b; Schaller 1963.
Orangutan	large	patchy	no, solitary	weak, males only	both sexes	no	solitary	Galdikas 1979; MacKinnon 1979; Rodman 1979.
Spider monkey	medium	patchy	yes	yes, males	females?	common	solitary or females	Cant 1978b.
Wild dog	medium	patchy	yes, during hunts	migratory, yes around den sites	females	some, during hunts	both sexes	Estes and Goddard 1967; Frame and Frame 1976; Kruuk and Turner 1967; Kuhme 1965; Schaller 1972.

Gray wolf	large	patchy	yes	yes, both sexes	both sexes	no	both sexes	Mech 1970, 1977a,b; Peters and Mech 1975.
Coyote	medium	patchy	yes	yes, variable	both sexes	no	both sexes	Bekoff and Wells 1980, 1981.
African lion	large	patchy	yes	yes, males	males primarily	yes, nomadic lions	both sexes	Bertram 1973, 1976; Rudnai 1973; Schaller 1972.
Spotted hyena	large	patchy	yes	yes	males	no	both sexes	Kruuk 1972.
Bottlenose dolphin	large	patchy	yes	?	?	common, subadults	females	Würsig 1979.
Whiptail wallaby	medium	"uniform" some coarse-grained patches	yes	yes, undefended "monopolized zones"	males	no	females	Kaufmann 1974.
Arabian babbler	small	patchy	yes?	yes	females	?	females	Zahavi 1974.

latrans), lions, and hyenas the patches are herds of, or sometimes individual, ungulates which wander throughout the habitat, or carrion (see table 37 for sources). Whiptail wallabies (*Macropus parryi*) split into small foraging parties to graze on scattered patches of grass in a habitat mosaic of varied types of vegetation (Kaufmann 1974). For Arabian Babblers (*Turdoides squamiceps*) patches consist of microclimates in which insects are found (Zahavi 1974). The nature of patches is not as relevant to the evolution of fusion-fission sociality as their abundance, distribution, and nutritional value in relation to the nutritional requirements and exploitative capabilities of the social group which preys upon them.

THE INFLUENCE OF FEMALE EXOGAMY

Exogamy influences the forms of cooperation within societies because of its influence on levels of genetic relatedness. Recently, Frame and Frame (1976) noted that, among mammals, only chimpanzees and wild dogs exhibited a pattern of female exogamy and male retention in social groups. Cant (1978b) described fusion-fission and cooperative defense of territory by male spider monkeys. From those data, plus the tendency for female spider monkeys to remain solitary and nonterritorial, it would not be surprising to learn that spider monkey societies are also male retentive and female exogamous. It is possible that continued research on bottlenose dolphins may yield a similar picture, although the society of these aquatic mammals may resemble that of lions, a species characterized by female retention and male exogamy (Schaller 1972; Bertram 1973). Arabian Babblers also exhibit male retention and female exogamy, and more significantly, males communally defend their territory. Fusion-fission societies exhibiting male retention and female exogamy are quite rare; all of them seem to be characterized by cooperative defense of the communal territory by males. It is difficult to postulate an evolutionary history for these male-retentive societies without invoking sexual selection and kin selection as the mechanisms which shaped them.

Although rare among mammals in general, male retentiveness and female exogamy is the most common mating and societal pattern among human hunter-gatherers (Murdock 1957: Thomas 1959; Hiatt 1968; Yengoyan 1968). Hunter-gatherers normally forage in small parties rather than as a discrete band, although males may hunt together when a communal technique is appropriate for the size and grouping tendencies of the prey (e.g., Hoebel 1960). Usually hunting occurs in small parties or solitarily (Turnbull 1961; Balikci 1968; Lee 1968; Woodburn 1968; Coon 1971 for a general review; Morgan 1979). In this way their patterns of resource exploitation resemble the normal pattern of fusion-fission societies. The communal defense of territory by male humans, which sometimes escalates to warfare (e.g., Vayda 1976), bears a striking similarity in basics to the territorial clashes between male chimpanzees in Gombe National Park as reported by Goodall (1979) and Goodall et al. (1979).

My intent here is not to attempt to represent chimpanzees as primitive humans, but in some ways they do present a human-like model. Humans are a successful species that have evolved social adaptations to exploit patchy and ephemeral resources in the ecologically sophisticated mode of hunter-gatherer. These adaptations include fusion-fission sociality, male retention, female exogamy, and cooperative defense of communal territories by males—a rare constellation of social and ecological adaptations evinced also by the species most closely related to humans, the chimpanzees.

Chimpanzees, as large-bodied, social specialists on fruit represent a rare situation in nature. Their fusion-fission social system and communal territoriality also are rare. The combination is so rare that were a biologist to propose the theoretical existence of a species with all of the attributes of chimpanzees to a primate ecologist, the latter probably would assure him on evolutionary grounds that his hypothetical creation was extremely improbable. If chimpanzees did not actually exist, we would consider it close to impossible that they could.

Much more remains to be learned about them. I have only scratched the surface of the social ecology of the chimpanzees of Kibale Forest. More data, of course, would clarify their behavior and reproductive strategies even more. It is my hope that such research continues. It is an even more dear hope to me that mankind will pause long enough in the present whirlwind of tropical deforestation to ensure the future survival of chimpanzees in their natural state.

APPENDIX A

Descriptions of individually known chimpanzees from Ngogo, Kibale Forest, Uganda as of May 1978. Age and sex class symbols are as follows: A, adult; SA, subadult; J, juvenile; I, infant; M, male; F, female; and ?, sex unknown. Immature chimpanzees whose names begin with the same first letter as the female above are assumed to be that female's offspring.

Name	Age-Sex Class	Size	Distinguishing Features and General Gestalt
Ardith	AF (middle-aged)	Large	Wavery indented margin, left ear only, large bodied, male size, heavy, old-looking face, sparsely haired ventrum, wide thick sideburns running in long, even arcs down sides of face. Pronounced downward dip to center of brow ridge.

Name	Age-Sex Class	Size	Distinguishing Features and General Gestalt
Ashly	JM (9 years)	Medium	Angular, hexagonal crown of head, well-developed musculature, pinkish-tan face and testicles (which are well-developed), short hair on crown, but long around fringe. Slight white tail tuft.
Anson	IM (3.5 years)	Large	Bare patch (6 cm²) lower sternum.
Blondie	AF (middle-aged)	Medium	Light brown to blonde tints all along dorsum and side burns, flatish crown, long sideburns, grayish muzzle with sharp chin demarcation. Handsome matron.
Bess	JF (7 years)	Medium	Ears widest at upper half, obvious, protruding clitoris, 5 cm long white tail tuft, dark tan face and ears.
Butch	IM (3 years)	Small	Looks "all head", 10 cm white tail tuft, average looking infant.
Clovis	AF (prime-middle-aged)	Medium	Two-colored dorsum teddy bear ears, short, thick head hair, flatish top to rounded crown, medium long sideburns, reddish skin tint below eyes into cheek, area, lower lip has slight pinkness at center, long hair on arms and body. Has wrinkled lips like Ruth Gordon.

Name	Age-Sex Class	Size	Distinguishing Features and General Gestalt
Clark	JM (7–8 years)	Medium	Punched-in, very gorilla-like nose, small-domed head, pink forehead bald spot, face, hands and feet. White tail tuft. *Ugly chimp*
Chita	IM (2.5–3 years)	Medium	—
Dumbo	AF (prime)	Medium	Two file-like notches in upper margin of right ear near to crown. Close notch is deep, 2nd notch is faint, looks like Disney character, Dumbo. Wide ears pointing outward at tops, face dark tan, wrinkled.
Eagle	AM (middle-aged)	Huge	Light brown hair on lower dorsum continuing in very light phase down thighs to calves. White beard, huge body, *massive musculature*. Medium length body hair. Head broad and massive, short hair, compact intact ears. Head appears to be sculpted in stone. Tan tint to face, especially around eyes. Unforgetably impressive.
Eskimo	AM (prime)	Large	Compact ears almost hidden from front view by very thick hair on sides of face and head. Like Eskimo parka hood. Flat profile to face. Body very large, well-developed musculature, long black

Name	Age-Sex Class	Size	Distinguishing Features and General Gestalt
Eskimo	AM (prime)	Large	hair on arms and shoulders, head massive and round.
Farkle	AF (middle-aged to old)	Small	Tired eyes, two toned dorsum, light hairs continuing onto thighs, longish face, shiney black forehead bald spot, rounded crown, head hair thick and moderately short brows close and well arched with pronounced center dip. Beard white, compact, intact teddy bear ears.
Fearless	SAM (12–13 years)	Small	Head hair very short, as is body hair, with traces of brown appearing on lower dorsum. Ears compact, left drops away from crown at 45° angle, right ear does not. Tan face, dark tan hands and feet, black hair. Testicles well-developed.
Felony	JF (8–9 years)	Medium	Ears entire, left ear has waver in upper half along margin, blotchy dark tan spots on pink-tan face, 10 cm long white tail tuft.
Fern	JF (5–6 years)	Medium	—
Fanny	IF (2 years)	Medium	—
Gray	AF (old)	Medium	Slight notch upper margin of left ear, right hand is missing forefinger, upper lip has vertical scar right of center, light grayish-

Name	Age-Sex Class	Size	Distinguishing Features and General Gestalt
Gray	AF (old)	Medium	brown tints to body hair, gray beard, flatish, wide, long-haired crown, compact ears, tired and retired matron.
(may be related to) Zira	AF (young, possibly subadult)	Medium	Lithe-limbed, svelte mover appears as a younger version of Gray. Long, thick sideburns, shoulder hair and body hair. Light muzzle not gray. Tan upper section of ears.
Zane	JF (5–6 years)	Medium	Left ear has slight mid-marginal undulation that right ear does not. Face is turning darker around muzzle in large, uneven blotches. Pinkish-tan face, ears, hands, and feet, white tail turf.
Hump	AF (old)	Medium	Hump prominently apparent in lateral sillouette of dorsum (sacral-lumbar region), broad full, rounded, black face reminiscent of television character, Archie Bunker. Two-tone tendency to lower back. Head and shoulder hair long.
Ita	AF (young, posibly subadult)	Small	Right arm is missing hand, left ear has radically-torn notch in upper margin, 10 cm² light patch right of sacrum, small bodied, black haired, face dark tan, head hair short.

Name	Age-Sex Class	Size	Distinguishing Features and General Gestalt
Joe	AM (young)	Medium	Left ear severely torn and healed with large gap in mid margin. Black back, dark tan face, moderately short head hair.
Kella	AF (young)	Small-Medium	Svelt (like primiparous female), body hair short, black head hair short, esp. crown, face pinkish-tan.
Kirk	IM (1–1.5 years)	Small	—
La	AF (young-prime)	Medium	Widely flaring ears, widest at top, prominent features. Body hair medium length, black not two-tone on dorsum, head hair short.
Lysa	IF (2–2.5 years)	Medium	Very long head and body hair (relative to other infants).
Mom	AF (prime to middle-age)		—
Mac	JM (6–7 years)		—
Munch	I? (1–1.5 years)		—
Nane	AM (prime)	Medium	Ears angular, with 45° slope away from crown, especially left ear. Notch in right ear near crown ~2.5 cm deep; face black, head hair moderately short lightish tints to whole of back, not two-toned.
Newman	SAM (verging on adult-hood)	Medium	Exaggerated pointed-domed cranium large "Gerald Ford" upper

Name	Age-Sex Class	Size	Distinguishing Features and General Gestalt
Newman	SAM (verging on adult-hood)	Medium	muzzle with an apparent scar running up center of upper lip ~4 cm. toward left side. Looks like *Mad* magazine cartoon character. Smooth wrinkles on ishial callosities-anal area is reminiscent of female. Well-developed testicles. Wide, intact, tan ears-widest at top
Notches	AM (middle-aged)	Large	Has a "bite in the apple" notch on the upper third (10:00 o'clock) of right ear *and* in the center (3:00 o'clock) of the left ear. Massive body, thick musculature. Thick facial hair, white beard, two-tone dorsum with light brown continuing down thighs.
Owl	AF (young, possibly subadult)	Small	Has notch in margin of right ear 3 cm deep, 2 cm from crown, head hair short, black sparse scraggly, body hair black, sparse, brown ridge peaked, not depressed, tan face, compact features, but homely.
Polly	AF (middle-aged)	Medium	Left ear has notch at 2:00 o'clock in rounded margin. Heavy bodied, black face and teats, ears compact, close to crown and evenly arced from frontal view.

Name	Age-Sex Class	Size	Distinguishing Features and General Gestalt
Punch	AF (sub-adult, young)	Small	Face has "pinched in from sides" effect (hour glass) when nose and muzzle join, head hair short, face tan, ears tan and very large brow ridge has center depression.
P——	I? (3 years)	Medium	—
Phantom	JM (7 years)	Medium	Facial hair very long, long ear tufts. Body stocky, round, longhaired face pinkish-tan turning blotchy brown, ears wide, smoothly arced. Left ear has marginal indentation. Brow ridge thickens, angling upward toward center then dips. Looks like Wolfman.
Quilla	AF (young-prime)	Small to Medium	Perineal swelling has pronounced hour glass shape with more hair and less pink than the average female. Compact ears, left ear shows pronounced mid-marginal waver, body and head hair moderately long, black, white beard. Wide arc and separation of brow ridges give her and exotic and esthetic look. A very attractive chimpanzee.
Raw Patch (R.P.)	SAM (verging on adulthood)	Medium	Has 7 cm diameter pink scar raw patch on lower left section of back. Head

Name	Age-Sex Class	Size	Distinguishing Features and General Gestalt
Raw Patch (R.P.)	SAM (verging on adulthood)	Medium	hair very short and ears very wide so that they wobble when he jerks his head. Head appears sculpted in stone, crown hair has brown highlights increasing toward nape to blond.
Satan	AM (middle-aged to old)	Medium to Large	Eyes tend to flash whitish, may have sclerotics. Second from outside toe on right foot is shorter than little toe, slight scar on upper lip 1 cm from center. Head hair short, crown rounded. Compact but wide teddy bear ears. Nostrils appear to flare, white beard with distinct chin boundary, double arched brows.
Shemp	AM (middle-aged)	Large	Right ear has midmarginal undulation. 3 cm scar running from right lower lip. Very long hair on arms, less so on shoulders. Very thick, long head hair, like parted in center. Massive body, hairs light-colored, increasing down back-not two-toned. Ears rounded, evenly arced, much hidden from frontal view by hair, deep set eyes, double-arched brows.

Name	Age-Sex Class	Size	Distinguishing Features and General Gestalt
Silverback (S.B.)	AM (prime to middle-aged)	Medium	Unusually sharp horizontal demarcation on two-tone back black & light brown. Small notch-like scar in midmargin of right ear, *difficult* to see. Black face, body, ears, hair, head hair thick, short on crown, thick round face, solid looking ears, compact. Massive musculature.
Stump	AM (old)	Large	Left arm is missing hand, body hair long, thick, especially on shoulders, arms and back. Brown-blond tints on lower back. Large head; black face framed by long hair. Eyes deep set, flash white when he glances upward. Face etched with wrinkles.
Spots	AM (middle-aged to old)	Medium	Gray beard with amelanic white spots at right corner of mouth. Lumpy ear margin, left ear visible from front. Bilaterally symmetrical two tone back, small toe on right foot is highly elongated and spike-like. Moderate build, not massive.

APPENDIX B

Descriptions of individually known chimpanzees from Kanyawara, Kibale Forest Reserve, Uganda as of April 1981. Age and sex class symbols are as follows: A, adult; SA, subadult; J, juvenile; I, infant; M, male; F, female; and ?, sex unknown. Immature chimpanzees whose names begin with the same first letter as the adult female above are assumed to be that female's offspring.

Name	Age-Sex Class	Size	Distinguishing Features and general Gestalt
Anka	AF (middle-aged)	large	Stocky and hunched body. Body hair, long, black, with slight two-tone dorsum. Pianist's fingers on left hand. Ears angle in at tops. Black face. Pink blotches rt. corner of mouth. Perineum appears almost separate when full.
Auk	JM (4–5 years)	small	No deformities.

Name	Age-Sex Class	Size	Distinguishing Features and General Gestalt
Baker	AF (prime+)	small,	Very light-colored hair, especially on lower dorsum and legs, two-tone back. Sparsely haired ventrum. Thin limbs, almost svelt. 5–8 cm dia. bald patch on left shoulder (difficult).
Brak	JM (~8 years)	medium	Small testicles. *Scar running from upper rt. foot (ankle) to shin. Has knoblike bump on lower rt. side of bare area (10–13 cm long) wide at base (difficult).
Beka	IF (2.5–3 years)	medium	—
Crack	AF (middle-aged) (lazy, moves little)	Medium-large	Short crown hair, almost bald. Body hair med., light brown on lower dorsum and thighs. Thin limbs and fingers. Intact ears. Hump on lower dorsum. Looks dopey.
Coke	J♀ (~6 years)		Homely. Large muzzle.
Darki	AF (middle-aged) (indifferent mother)	small-medium	Mostly black hair, brown lower dorsum, long. Like juvenile. Dark tan face. small notch upper ⅓ of left ear.
Donika	JF (8 years)	medium	Short black crown hair. Light tan face. **Right hand partially crippled; snare wire polished and embedded distal to wrist. Limb cannot bear weight or manipulate; hand dangles limply.

Name	Age-Sex Class	Size	Distinguishing Features and General Gestalt
Derek	IM (≥3 years)	medium	—
Hook	AM (middle-aged)	medium	Massive musculature. Deep furrow in crown hair. Compact, crescent ears. Black face. Long black crown hair. Silver dollar sized bald patch scar above rt. temple. **Left hand is missing all of pinkie and parts of other 3 fingers, which are crippled & spider-thin.
Ikarus	AM (prime-middle-aged)	small-medium	Moderate musculature. Long body hair, 2-tone dorsum. Black face, light brown eyes. **Rt. hand mutilated; thumb or forefinger appears intact; other fingers are gone. *Bald scar rt. forehead.
Jack	AM (prime)	medium	Heavy musculature. Tan face. Short crown hair. Ears wide at tops, partly hidden by long hair. Looks like bruiser. **2nd finger on left hand is 1 digit short, odd knuckle callous pattern. **3rd toe 1. foot is nub only.
Kalhoun	AM (prime)	large	Massive musculature. Long body hair. Tan-pinkish face.
King	AM (old)	medium-large	Well-developed musculature. Huge testicles. Short-med. length body hair, very light colored, 2-tone

Name	Age-Sex Class	Size	Distinguishing Features and General Gestalt
King	AM (old)	medium-large	dorsum ext. down thighs & calves. Crown hair light, medium length, thin. Gray beard, white edges. Compact, wrinkled, lightly pigmented face. Fingers and toes intact.
Klubfoot	AM (middle-aged)	medium	Light, thin musculature. Body hair medium, light, 2-tone tendency on dorsum. **Left ear torn and healed visible in left profile only. **Right foot crippled and club-like, twisted *outward* and does not grasp. Right leg partially atrophied.
Knuck	SAM (~11 years)		Light musculature, female-like. Tan face. No upper canines. *Thin 2.5 cm scar running down from left corner of mouth (difficult). *1st knuckle on 2nd fin-of rt. hand enlarged in profile only (difficult)
Kong	AM (prime) (dominant individual)	medium	Well-developed musculature. Body hair black, *slight* brown dorsum, long, thick. Striking light tan face, white beard, black hair on crown. High-domed saggital area (gorilla). All fingers & toes o.k.
Veek	AF (middle-aged) (indifferent mother)	large	Rotund, black body hair, long, 2-tone tendency on dorsum. Thick black hair

Name	Age-Sex Class	Size	Distinguishing Features and General Gestalt
Veek	AF (middle-aged) (indifferent mother)	large	on crown. *Narrow ∧ - shaped forehead bald patch. Small scar on left temple (?) *2.5 cm knob 10 cm up from right heel, behind.
Valkerie	JF (~8 years)	medium	Dark tan face, cute and well-proportioned. All digits intact. Short hair on crown.
Vik	IM (~3 years)	medium	Homely face.

REFERENCES

Albrecht, H. 1976. Chimpanzees in Uganda. *Oryx* 13:357–361.

Albrecht, H. and S. C. Dunnet. 1971. *Chimpanzees in Western Africa*. Munich: R. Piper and Co.

Altmann, S. A. 1974. Baboons, space, time and energy. *Amer. Zool.* 14:221–248.

Asdell, S. A. 1946. *Patterns of Mammalian Reproduction*. New York: Comstock. (Cited by Goodall 1968.)

Azuma, S. and A. Toyoshima. 1962. Progress report of the survey of chimpanzees in their natural habitat, Kabogo Point area, Tanganyika. *Primates* 3:61–70.

Balikci, A. 1968. The Netsilik Eskimos: Adaptive processes. In R. B. Lee and I. DeVore, eds. *Man the Hunter*, pp. 30–48. Chicago: Aldine.

Bauer, H. R. 1979. Agonistic and grooming behavior in the reunion context of Gombe Stream chimpanzees. In Hamburg and McCown 1979:394–403.

Bekoff, M. and M. C. Wells. 1980. The social ecology of coyotes. *Scientific American* 242 (4):130–148.

Bekoff, M. and M. C. Wells. 1981. Behavioural budgeting by wild coyotes: The influence of food resources and social organization. *Anim. Behav.* 29:794–801.

Bell, R. H. V. 1971. A grazing ecosystem in the Serengeti. *Scientific American* 225 (1):86–93.

Bertram, B. C. R. 1973. Lion population regulation. *E. Afr. Wild. J.* 11:215–225.

Bertram, B. C. R. 1976. Kin selection in lions and in evolution. In P. P. G. Bateson and R. Hinde, eds. *Growing Points in Ethology*, pp. 281–301. Cambridge: Cambridge University Press.

Bolwig, N. 1959. A study of nests built by mountain gorilla and chimpanzee. *S. Afr. J. Sci.* 55:286–291.

Brody, S. 1945. *Bioenergetics and Growth*. New York: Hafner. (Cited by Rodman 1979).

Brooks, G. R. Jr. 1967. Population ecology of the ground skink *Lygosoma laterale* (Say). *Ecol. Monogr.* 37:71–89.

Brown, J. L. 1969. Territorial behavior and population regulation among birds. *Wilson Bull.* 81:293–329.

Buechner, H. K. 1971. Implications of social behavior in management of the Uganda kob. In *The behavior of ungulates and its relation to management.* Vol. 2, Alberta, Canada: ICUN Conference, Univ. of Calgary.

Busse, C. D. 1977. Chimpanzee predation as a possible factor in the evolution of red colobus monkey social organization. *Evolution* 31:907–911.

Busse, C. D. 1978. Do chimpanzees hunt cooperatively? *Am. Nat.* 112:767–770.

Bygott, J. D. 1972. Cannibalism among wild chimpanzees. *Nature* 238:410–411.

Bygott, J. D. 1979. Agonistic behavior, dominance, and social structure in wild chimpanzees of the Gombe National Park. In Hamburg and McCown 1979:404–427.

Cahill, T. 1981. Gorilla Tactics, life and love in gorilla country. *Geo* 3(12):100–116.

Cant, J. G. H. 1978a. Population survey of spider monkeys (*Ateles geoffroyi*) at Tikal, Guatemala. *Primates* 19:525–535.

Cant, J. G. H. 1978b. Ecology, locomotion, and social organization of spider monkeys (*Ateles geoffroyi*). Ph.D. dissertation, University of California, Davis.

Carrick, R. 1963. Ecological significance of territory in the Australian magpie. *Proc. XIII Intern. Ornithol. Congr.*:740–753.

Casimir, M. J. 1979. An analysis of gorilla nesting sites of the Kahuzi Region (Zaire). *Folia Primatol.* 32:290–308.

Clark, C. B. 1977. A preliminary report on weaning among chimpanzees of the Gombe National Park, Tanzania. In S. Chevalier-Skolnikoff and F. E. Porier, eds. *Primate Bio-Social Development: Biological, Social, and Ecological Determinants*, pp. 235–260. New York: Garland.

Clutton-Brock, T. H. 1972. Feeding and ranging behavior of the red colobus monkey. Ph.D. dissertation, University of Cambridge.

Clutton-Brock, T. H. 1974. Primate social organization and ecology. *Nature* 250:539–542.

Clutton-Brock, T. H. and P. H. Harvey. 1977. Species differences in feeding and ranging behavior in primates. In T. H. Clutton-Brock, ed. *Primate Ecology: Studies of Feeding and Ranging Behavior in Lemurs, Monkeys and Apes*, pp. 557–584. London: Academic Press.

Cody, M. L. 1966. A general theory of clutch size. *Evolution* 20:174–184.

Cole, L. C. 1954. The population consequences of life history phenomena. *Quart. Rev. Biol.* 29:103–137.

Committee on Injuries, American Academy of Orthopedic Surgeons. 1971. *Emergency Care and Transportation of the Sick and Injured.* Chicago: American Academy of Orthopedic Surgeons.

Coon, C. S. 1971. *The Hunting Peoples.* Boston: Little Brown, and Co.

Cowles, J. T. 1937. Food tokens as incentives for learning by chimpanzees. *Phychol. Monogr.* 14:1–88.

Crook, J. H. 1970. The socio-ecology of primates. In J. H. Crook, ed. *Social Behavior in Birds and Mammals*, pp. 103–166. New York: Academic Press.

Crook, J. H. and J. S. Gartlan. 1966. Evolution of primate societies. *Nature* 210:1200–1203.

Darwin, C. 1859. *The Origin of Species.* London: John Murray.

Darwin, C. 1871. *The Descent of Man, and Selection in Relation to Sex.* London: John Murray.

Dawkins, R. 1976. *The Selfish Gene*. Oxford: Oxford University Press.

Dice, L. R. 1952. *Natural Communities*. Ann Arbor: University of Michigan Press. (cited by Brooks 1967.)

Dobzhansky, T. 1962. *Mankind Evolving*. New Haven: Yale University Press.

Dorst, J. and P. Dandelot. 1970. *A Field Guide to the Larger Mammals of Africa*. London: Collins.

Eggeling, W. J. and I. R. Dale. 1951. *The Indigenous Trees of the Uganda Protectorate*. Uganda Protectorate: Government Printer.

Eisenberg, J. F., N. A. Muckenhirn, and R. Rudran. 1972. The relation between ecology and social structure in primates. *Science* 176:863–874.

Eisenberg, J. F. and R. W. Thorington Jr. 1973. A preliminary analysis of a neotropical fauna. *Biotropica*. 5(3):150–161.

Ellefson, J. O. 1968. Territorial behavior in the common white-handed gibbon, *Hylobates lar* Linn. In P. C. Jay, ed. *Primates Studies in Adaptation and Variability*, pp. 180–199. San Francisco: Holt, Rinehart and Winston.

Elton, C. and R. S. Miller. 1954. The ecological survey of animal communities: With a practical system of classifying habitats by structural characters. *J. Ecol.* 42:480–496.

Eltringham, S. K. and R. C. Malpas. 1976. Elephant slaughter in Uganda. *Oryx* 13:334–335.

Emlen, J. M. 1966. The role of time and energy in food preference. *Am. Nat.* 100:611–617.

Emlen, J. M. 1973. *Ecology: An Evolutionary Approach*. Menlo Park, Ca.: Addison-Wesley.

Emlen, J. T. 1971. Population densities of birds derived from transect counts. *The Auk* 88:323–342.

Estes, R. and J. Goddard. 1967. Prey selection and hunting behavior of the African wild dog. *J. Wild. Mgmt.* 31:52–70.

Ferster, C. B. 1964. Arithmetic behavior in chimpanzees. *Scientific American* 210(2):98–106.

Fossey, D. 1970. Making friends with mountain gorillas. *National Geographic* 137:48–68.

Fossey, D. 1971. More years with mountain gorillas. *National Geographic* 140:574–585.

Fossey, D. 1974. Observations on the home range of one group of mountain gorillas (*Gorilla gorilla berengei*). *Anim. Behav.* 22:568–581.

Fossey, D. 1976. The great apes, a dialogue with Jane Goodall, Dian Fossey, Birute Galdikas-Brindamour. The L. S. B. Leakey Foundation News, No. 6.

Fossey, D. 1979. Development of the mountain gorilla (*Gorilla gorilla beringei*): The first thirty-six months. In Hamburg and McCown 1979:139–186.

Fossey, D. 1981. Imperiled giants of the forest. *National Geographic* 159:501–523.

Fossey, D. 1982. Queen of the apes. *Scope* 17(23):48–59.

Fouts, R. S. and R. L. Budd. 1979. Artificial and human language acquisition in the chimpanzee. In Hamburg and McCown 1979:374–92.

Frame, L. H. and G. W. Frame. 1976. Female African wild dogs emigrate. *Nature* 263:227–229.

Freeland, W. J. 1977. Blood sucking flies and primate polyspecific associations. *Nature* 269:801–802.

Frisch, J. E., S.J. 1968. Individual behavior and intertroop variability in Japanese macaques. In P. C. Jay, ed. *Primates Studies in Adaptation and Variability*, pp. 243–252. San Francisco: Holt, Rinehart and Winston.

213 **REFERENCES**

Gadgil, M. 1972. Male dimorphism as a consequence of sexual selection. *Am. Nat.* 106:574–580.

Gadgil, M. and W. H. Bossert. 1970. Life historical consequences of natural selection. *Am. Nat.* 104:1–24.

Galdikas, B. M. F. 1979. Orangutan adaptation at Tanjung Puting Reserve: Mating and ecology. In Hamburg and McCown 1979:195–234.

Gartlan, J. S. and T. T. Struhsaker. 1972. Polyspecific associations and niche separation of rainforest anthropoids in Cameroon, West Africa. *J. Zool., Lond.* 168:221–266.

Gaulin, S. J. C. 1979. A Jarman/Bell model of primate feeding niches. *Human Ecol.* 7(1):1–20.

Gautier, J. P. and A. Gautier-Hion. 1969. Les associations polyspecifiques chez les Cercopithecidae du Gabon. *Terre et la Vie* 2:164–201.

Gautier-Hion, A. and J. P. Gautier. 1974. Les associations polyspecifiques de Cercopithques du plateau du M'passa (Gabon). *Folia primatol.* 22:134–177.

Geist, V. 1974. On the relationship of social evolution and ecology in ungulates. *Amer. Zool.* 14(1):205–220.

Glass, H. B. 1953. The genetics of the dunkers. *Scientific American* 189(2):76–81.

Goodall, J. 1963. My life among wild chimpanzees. *National Geographic* 124:272–308.

Goodall, J. 1965. Chimpanzees of the Gombe Stream Reserve. In I. DeVore, ed. *Primate Behavior Field Studies of Monkeys and Apes,* pp. 425–473. San Francisco: Holt, Rinehart and Winston.

Goodall, J. v. L. 1968. The behaviour of free-living chimpanzees in the Gombe Stream Reserve. *Anim. Behav. Monogr.* 1:161–311.

Goodall, J. v. L. 1971a. *In the Shadow of Man.* Boston: Houghton Mifflin.

Goodall, J. v. L. 1971b. Some aspects of aggressive behavior in a group of free-living chimpanzees. *Int. Soc. Sci. J.* 23:89–97.

Goodall, J. v. L. 1973a. The behavior of chimpanzees in their natural habitat. *Amer. J. Psychiatry* 130:1–12.

Goodall, J. v. L. 1973b. Cultural elements in a chimpanzee community. *Symp. IVth Int. Congr. Primat.,* Vol. 1:144–184.

Goodall, J. v. L. 1977. Infant killing and cannibalism in free-living chimpanzees. *Folia primatol.* 28:259–282.

Goodall, J. 1979. Life and death at Gombe. *National Geographic* 155:592–621.

Goodall, J., A. Bandoro, E. Bergmann, C. Busse, H. Matama, E. Mpongo, A. Pierce, and D. Riss. 1979. Intercommunity interactions in the chimpanzee population of the Gombe National Park. In Hamburg and McCown 1979:13–54.

Goss-Custard, J. D., R. I. M. Dunbar, and F. P. G. Aldrich-Blake. 1972. Survival, mating, and rearing strategies in the evolution of primate social structure. *Folia primatol.* 17:1–19.

Graham, C. F. and H. M. McClure. 1977. Ovarian tumors and related lesions in aged chimpanzees. *Veterinary Pathology* 14:380–386.

Grand, T. 1972. A mechanical interpretation of terminal branch feeding. *J. Mammal.* 53:198–201.

Grossman, M. L. and J. Hamlet. 1964. *Birds of Prey of the World.* New York: Bonanza.

Gunther, M. 1971. *Infant feeding.* Harmondsworth, England: Penguin.

Halperin, S. D. 1979. Temporary association patterns in free ranging chimpanzees. In Hamburg and McCown 1979:490–499.

Hamburg, D. A. and E. R. McCown. 1979. *The Great Apes.* Menlo Park, Calif.: Benjamin/Cummings.

Hamilton, W. D. 1964. The genetical theory of social behavior, I, II. *J. of Theoret. Biol.* 12:1–52.

Hamilton, W. D. 1971. Geometry for the selfish herd. *J. of Theoret. Biol.* 31:295–311.

Hamilton, W. D. 1972. Altruism and related phenomena, mainly in social insects. *Ann. Rev. of Ecol. and System.* 3:193–232.

Hamilton, W. J., III. and C. D. Busse. 1978. Primate carnivory and its significance to human diets. *BioScience* 28:761–766.

Harcourt, A. H. 1978. Strategies of immigration and transfer by primates, with particular reference to gorillas. *Z. Tierpsychol.* 48:401–420.

Harcourt, A. H. 1979a. The social relations and group structure of wild mountain gorilla. In Hamburg and McCown 1979:186–192.

Harcourt, A. H. 1979b. Contrasts between male relationships in wild gorilla groups. *Behavioral Ecology and Sociobiology* 5(1):39–50.

Heindrichs, H. 1970. Schatzungen der huptierbiomass in der dorbushsavanne nordlich und weslich der Serengetisteppe in Ostafrica nach einem neuen verfaren und bemerkungen zur biomass der anderen pflanzen-fressenden tierarten. *Saugetierk. Mitt.* 18(3):237–255. (cited by Schaller 1972).

Hiatt, L. R. 1968. Gidjingali marriage arrangements. in R. B. Lee and I. DeVore, eds., *Man the Hunter*, pp. 165–175. Chicago: Aldine.

Hinde, R. A. 1956. The biological significance of territories in birds. *Ibis* 98:340–369.

Hladik, C. M. 1977. Chimpanzees of Gabon and chimpanzees of Gombe: Some comparative data on the diet. in T. H. Clutton-Brock, ed. *Primate Ecology: Studies of Feeding and Ranging Behavior in Lemurs, Monkeys and Apes*, pp. 481–501. London: Academic Press.

Hoebel, E. A. 1960. *The Cheyenne Indians of the Great Plains.* San Francisco: Holt, Rinehart and Winston.

Horn, H. S. 1968. The adaptive significance of colonial nesting in the Brewer's Blackbird (*Euphagus cyanocephalus*). *Ecology* 49:682–694.

Hrdy, S. B. 1977. Infanticide as a primate reproductive strategy. *Amer. Scient.* 65:40–49.

Hrdy, S. B. 1979. Infanticide among animals: A review, classification, and examination of the implications for the reproductive strategies of females. *Ethology and Sociobiology* 1(1):13–40.

Hubbell, S. 1979. Tree dispersion, abundance, and diversity in a dry tropical forest. *Science* 203:1299–1309.

Hutchins, M. and D. P. Barash. 1976. Grooming in primates: Implications for its utilitarian function. *Primates* 17(2):145–150.

Hutchinson, G. E. 1957. Concluding remarks. *Cold Spring Harbor Symp. Quant. Biol.* 22:198–201.

Huxley, T. H. 1863. *Evidence as to Man's Place in Nature.* London: William and Norgate.

Itani, J. and A. Suzuki. 1967. The social unit of chimpanzees. *Primates* 8:355–381.

Izawa, K. 1970. Unit groups of chimpanzees and their nomadism in the savanna woodland. *Primates* 11:1–46.

Izawa, K. and J. Itani. 1966. Chimpanzees in Kasakati Basin, Tanganyika (I). Ecological study in the rainy season 1963–1964. *Kyoto Univ. Afr. Stud.* 1:73–156.

Janson, C. and J. Terborgh. No date. Censusing primates in rainforest. Princeton, N.J.: Dept. of Biology, Princeton University.

Janzen, D. H. 1979. How to be a fig. *Ann. Rev. Ecol. Syst.* 10:13–52.

Jarman, P. J. 1974. The social organization of antelope in relation to their ecology. *Behavior* 58(3,4):215–267. (cited by Gaulin 1979).

Jolly, A. 1972. *The Evolution of Primate Behavior*. N ew York: MacMillan.

Jones, C. and J. Sabater Pi. 1971. Comparative ecology of *Gorilla gorilla* (Savage and Wyman) and *Pan troglodytes* (Blumenback) in Rio Muni, West Africa. *Bibliotheca Primatologica*.

Joslin, P. W. B. 1966. Summer activities of two timber wolf (*Canis lupus*) packs in Algonquin Park. M. S. thesis, Univ. of Toront. (Cited by Mech 1970.)

Joubert, E. 1972. The social organization and associated behaviour in the Hartmann zebra, *Equus zebra hartmannae*. *Madoqua* 1(6):17–56.

Kano, T. 1971. The chimpanzees of Filabanga, Western Tanzania. *Primates* 12:229–246.

Kano, T. 1972. Distribution and adaptation of the chimpanzee on the eastern shore of Lake Tanganyika. *Kyoto Univ. African Studies* 7:37–129.

Kano, T. 1980. Social behavior of wild pygmy chimpanzees (*Pan paniscus*) of Wamba: A preliminary report. *J. Human Evol.* 9:243–260.

Kaufmann, J. H. 1974. Social ethology of the whiptail wallaby, *Macropus parryi*, in northeastern New South Wales. *Anim. Behav.* 22:281–369.

Kawabe, M. 1966. One observed case of hunting behavior among wild chimpanzees living in the savanna woodland of western Tanzania. *Primates* 7:393–396.

King, M. C. and A. C. Wilson. 1975. Evolution at two levels in humans and chimpanzees. *Science* 188:107–116.

Kingston, B. 1967. *Working Plan for Kibale and Itwara Central Forest Reserves*. Entebbe, Uganda: Forestry Department.

Klein, L. L. and D. J. Klein. 1973. Observations on two types of neotropical primate intertaxa associations. *Am. J. Phys. Anthrop.* 38:649–654.

Kohler, W. 1925. *The Mentality of Apes*. London: Routledge and Kegan Paul.

Kormondy, E. J. 1969. *Concepts of Ecology*. Englewood Cliffs, New Jersey: Prentice-Hall.

Kortlandt, A. 1962. Chimpanzees in the wild. *Scientific American* 206(5):128–138.

Kruuk, H. 1972. *The Spotted Hyena*. Chicago: University of Chicago Press.

Kruuk, H. and M. Turner. 1967. Comparative notes on predation by lion, leopard, cheetah, and wild dog in the Serengeti area, East Africa. *Mammalia* 31:1–27.

Kuhme, W. 1965. Communal food distribution and division of labor in African hunting dogs. *Nature* 205:443–444.

Kuhn, H. J. 1968. Parasites and the phylogeny of the catarrhine primates. in B. Chiarelli, ed. *Taxonomy and Phylogeny of Old World Primates with References to the Origin of Man*, pp. 187–195. Turin: Rosenberg and Sellier. (cited by Hutchins and Barash 1976.)

Kummer, H. 1971. Immediate causes of primate social structures. *Proc. 3rd. Int. Cong. Primat., Zurich* 3:1–11.

Lee R. B. 1968. What hunters do for a living, or, how to make out on scarce resources. In R. B. Lee and I. DeVore, eds. *Man the Hunter*, pp. 30–48. Chicago: Aldine.

MacArthur, R. H. and E. R. Pianka. 1966. On optimal use of a patchy environment. *Am. Nat.* 100:603–609.

MacKinnon, J. 1974. The behavior and ecology of wild orangutans *Pongo pygmaeus*. *Anim. Behav.* 22:3–74.

MacKinnon, J. 1979. Reproductive behavior in wild orangutan populations. In Hamburg and McCown 1979:256–273.

Marler, P. 1969. Vocalizations of wild chimpanzees. *Proc. 2nd Int. Congr. Primat.* 1:194–100.

Marler, P. and L. Hobbet. 1975. Individuality in a long-range vocalization of wild chimpanzees. *Z. Tierpsychol.* 38:97–109.

Mason, W. A. 1968. Use of space by *Callicebus* groups. in P. C. Jay, ed. *Primates: Studies in Adaptation and Variability*, pp. 200–216. San Francisco: Holt, Rinehart and Winston.

Mason, W. A. 1970. Chimpanzee social behavior. in G. H. Bourne, ed. *The Chimpanzee*, 2:265–288. New York: Karger.

McGinnis, P. R. 1979. Sexual behavior in free-living chimpanzees: Consort relationships. In Hamburg and McCown 1979:428–439.

McGrew, W. C. 1975. Patterns of plant food sharing by wild chimpanzees. in *Proc. Vth Congr. Int. Primatol. Soc. Nagoya, Japan*, pp. 304–309.

McGrew, W. C. 1977. Socialization and object manipulation of wild chimpanzees. in S. Chevalier-Skolnikoff and F. E. Poirier, eds., *Biosocial Development: Biological, Social, and Ecological determinants*, pp. 261–288. New York: Garland.

McGrew, W. C. 1979. Evolutionary implications of sex differences in chimpanzee predation and tool use. In Hamburg and McCown, 1979:440–463.

McGrew, W. C. and C. E. G. Tutin. 1978. Evidence for a social custom in wild chimpanzees? *Man* 13:234–251.

McKey, D., P. G. Waterman, C. N. Mbi, J. S. Gartlan, and T. T. Struhsaker. 1978. Phenolic content of vegetation in two African rainforests: Ecological implications. *Science* 202:61–64.

McNab, B. 1963. Bioenergetics and the determination of home range size. *Am. Nat.* 97:133–140.

Mech, L. D. 1970. *The Wolf: The Ecology and Behavior of an Endangered Species.* New York: Natural History Press.

Mech, L. D. 1977a. Productivity, mortality, and population trends of wolves in northeastern Minnesota. *J. Mammal.* 58:559–574.

Mech, L. D. 1977b. Where can the wolf survive? *National Geographic* 152:518–537.

Menzel, E. W., Jr. 1973. Chimpanzee spatial memory organization. *Science* 182:943–945.

Menzel, E. W. Jr. 1979. Communication of object locations in a group of young chimpanzees. In Hamburg and McCown 1979:358–371.

Merfield, F. G. and H. Miller. 1956. *Gorillas Were my Neighbors.* London: Longmans.

Miller, D. A. 1977. Evolution of primate chromosomes. *Science* 198:1116–1124.

Miller, R. S. 1967. Pattern and process in competition. *Adv. Ecol. Res.* 4:1–74.

Milton, K. and M. L. May. 1976. Body weight, diet and home range area in primates. *Nature* 259:459–462.

Morgan, C. J. 1979. Eskimo hunting groups, social kinship, and the possibility of kin selection in humans. *Ethology and Sociobiology* 1(1):83–86.

Morris, K. and J. Goodall. 1977. Competition for meat between chimpanzees and baboons of the Gombe National Park. *Folia primatol.* 28:109–121.

Murdock, G. P. 1957. World ethnographic sample. *American Anthropologist* 59:664–687.

Murie, A. 1944. *The Wolves of Mount McKinley.* U.S. Nat. Park Service Fauna Ser. #5 (Government Printing Office, Washington, D. C.).

Nicholson, A. J. 1957. Self-adjustment of population to change. *Cold Spring Harbor Symp. Quant. Biol.* 22:153–173.

Nishida, T. 1968. The social group of wild chimpanzees in the Mahali Mountains. *Primates* 9:167–224.

Nishida, T. 1970. Social behavior and relationship among wild chimpanzees of the Mahali Mountains. *Primates* 11:47–87.

Nishida, T. 1973. The ant-gathering behaviour by the use of tools among wild chimpanzees of the Mahali Mountains. *J. Hum. Evol.* 2:357–370.

Nishida, T. 1979. The social structure of chimpanzees of the Mahale Mountains. In Hamburg and McCown 1979:73–122.

Nishida, T. and K. Kawanaka. 1972. Inter-unit-group relationships among wild chimpanzees of the Mahali Mountains. *Kyoto Univ. Afr. Stud.* 7:131–169.

Nissen, N. W. 1931. A field study of the chimpanzee. *Comp. Psychol. Monogr.* 8:1–122.

Oates, J. F. 1974. The ecology and behavior of black and white colobus (*Colobus guereza* Ruppel) in East Africa. Ph.D. dissertation, University of London.

Oosting, H. J. 1956. *The Study of Plant Communities* San Francisco: W. H. Freeman.

Osmaston, H. A. 1959. *Working plan for Kibale and Itwara Forests* (First Rev.). Entebbe, Uganda: Uganda Forestry Department. (Cited by Wing and Buss 1970.)

Park, T. 1954. Experimental studies of interspecies competition. II. Temperature, humidity, and competition in two species of *Tribolium*. *Physiol. Zool.* 27:177–238.

Peters, R. P. and L. D. Mech. 1975. Scent-marking in wolves. *Amer. Scient.* 63:628–637.

Peterson, R. S. and G. A. Bartholomew. 1967. The natural history and behavior of the California sea lion. *Amer. Soc. Mammalogists, Spec. Publ.* No. 1.

Pierce, A. H. 1978. Ranging patterns and associations of a small community of chimpanzees in Gombe National Park, Tanzania. In D. J. Chivers and J. Herbert, eds. *Recent advances in primatology volume 1 behavior*, pp. 59–61. San Francisco: Academic Press.

Platt, J. R. 1964. Strong inference. *Science* 146:347–353.

Popp, J. L. and I. DeVore. 1979. Aggressive competition and social dominance theory: Synopsis. In Hamburg and McCown 1979:316–338.

Pusey, A. 1979. Intercommunity transfer of chimpanzees in Gombe National Park. In Hamburg and McCown 1979:465–480.

Reid, R. M. 1976. Effects of consanguineous mating and inbreeding on couple fertility and offspring mortality in rural Sri Lanka. In B. A. Kaplan, ed. *Anthropological studies of human fertility*, pp. 139–146. Detroit: Wayne State Univ. Press.

Reynolds, V. 1963. An outline of the behavior and social organization of forest-living chimpanzees. *Folia primatol.* 1:95–102.

Reynolds, V. 1975. How wild are the Gombe chimpanzees? *Man* 10:123–125.

Reynolds, V. and F. Reynolds. 1965. Chimpanzees in the Budongo Forest. In I. DeVore, ed. *Primate Behavior: Field Studies of Monkeys and Apes*, pp. 468–524. San Francisco: Holt, Rinehart and Winston.

Richards, P. W. 1952. *The Tropical Rain Forest*. London: Cambridge University Press.

Rijksen, H. D. 1978. *A Field Study on Sumatran Orangutans (Pongo pygmaeus abelii Lesson 1927): Ecology, Behavior and Conservation*. Wageningen, the Netherlands: H. Veenman and Zonen, B. V.

Riss, D. C. and C. R. Busse. 1977. Fifty-day observation of a free-ranging adult male chimpanzee. *Folia primatol.* 28:283–297.

Riss, D. C. and J. Goodall. 1977. The recent rise to alpha-rank in a population of free-living chimpanzees. *Folia primatol.* 27:134–151.

Robbins, D. and C. T. Bush. 1973. Memory in great apes. *J. Exp. Psychol.* 97:344–348.

Robinette, W. L., C. M. Loveless, and D. A. Jones. 1974. Field tests of strip census methods. *J. Wildlf. Mgmt.* 38:81–96.

Rodman, P. S. 1973a. Population composition and adaptive organization among orangutans of the Kutai Reserve. In R. P. Michael and J. H. Crook, eds. *Ecology and Behavior of Primates*, pp. 171–209. London: Academic Press.

Rodman, P. S. 1973b. Synecology of Bornean primates. Ph.D. dissertation, Harvard University, Cambridge, Mass.

Rodman, P. S. 1978. Diets, density and distributions of Bornean primates. In G. G. Montgomery, ed. *Ecology of Arboreal Folivores*, pp. 465–478. Washington, D. C.: Smithsonian Press.

Rodman, P. S. 1979. Individual activity patterns and the solitary nature of orangutans. In Hamburg and McCown 1979:235–256.

Rowell, T. E. 1966. Forest living baboons in Uganda. *J. Zool., Lond.* 149:344–364.

Ruch, T. C. 1959. Diseases of the skin. In T. C. Ruch, ed. *Diseases of Laboratory Primates*, pp. 501–528. Philadelphia: Saunders.

Rudnai, J. 1973. Reproductive biology of lions (*Panthera leo massaica* Neumann) in Nairobi National Park. *E. Afr. Wildl. J.* 11:241–253.

Rudran, R. 1976. The socio-ecology of the blue monkey (*Cercopithecus mitis Stuhlmanni*) of the Kibale Forest, Uganda. Ph.D. disseration, Univ. of Maryland.

Rumbaugh, D. M. 1970. Learning skills of anthropoids. In L. A. Rosenblum, ed. *Primate Behavior*, 1:1–70. London: Academic Press.

Sarich, V. M. and J. E. Cronin. 1976. Molecular systematics of the primates. In M. Goodman and R. Tashian, eds. *Molecular Anthropology*. New York: Plenum. (Cited by Hamburg and McCown 1979.)

Sayfarth, R. M. 1976. Social relationships among adult female baboons, *Anim. Behav.* 24:917–928.

Schaller, G. B. 1961. The orang-utan in Sarawak. *Zoologica* 46:73–82.

Schaller, G. B. 1963. *The Mountain Gorilla*. Chicago: University of Chicago Press.

Schaller, G. B. 1972. *The Serengeti Lion*. Chicago: University of Chicago Press.

Schoener, T. 1968. Sizes of feeding territories among birds. *Ecology* 49:123–141.

Schoener. T. 1971. Theory of feeding strategies. *Ann. Rev. Ecol. Systemat.* 2:369–404.

Short, R. V. 1979. Sexual selection and its component parts, somatic and genital selection, as illustrated by man and the great apes. *Advances in the Study of Behavior*, 9:131–158.

Silk, J. B. 1978. Patterns of food sharing among mother and infant chimpanzees at Gombe National Park, Tanzania. *Folia primatol.* 29:129–141.

Simpson, M. J. A. 1973. The social grooming of male chimpanzees. In R. P. Michael and J. H. Crook, eds. *Comparative Ecology and Behaviour of Primates*. pp. 411–505. New York: Academic Press.

Smith, C. C. 1968. The adaptive nature of social organization in the genus of tree squirrels *Tamiasciurus. Ecol. Monogr.* 38:31–63.

Smith, C. C. 1977. Feeding behavior and social organization in howling monkeys. In T. H. Clutton-Brock, ed. *Primate Ecology: Studies of Feeding and Ranging Behavior in Lemurs, Monkeys and Apes*, pp. 97–126. London: Academic Press.

Sparks, J. 1967. Allogrooming in primates: A review. In D. Morris, ed. *Primate Ethology*, pp. 190–225. Garden City, N. Y.: Doubleday. (Cited by Goodall 1968.)

Struhsaker, T. T. 1967. Social structure among vervet monkeys (*Cercopithecus aethiops*). *Behaviour* 29:83–121.

Struhsaker, T. T. 1969. Correlates of ecology and social organization among African cercopithecines. *Folia primatol.* 11:80–118.

Struhsaker, T. T. 1975. *The Red Colobus Monkey*. Chicago: University of Chicago Press.

Struhsaker, T. T. 1977. Infanticide and social organization in the redtail monkey (*Cercopithecus ascanius schmidti*) in the Kibale Forest, Uganda. *Z. Tierpsychol.* 45:75–84.

Struhsaker, T. T. 1978. Food habits of five monkey species in the Kibale Forest, Uganda.

In D. C. Chivers and J. Herbert. eds. *Recent Advances in Primatology*, 1:225–248. London: Academic Press.

Struhsaker, T. T. 1980. Comparison of the behavior and ecology of redtail and red colobus monkeys in the Kibale Forest, Uganda. *Afr. J. Ecol.* 18:33–51.

Struhsaker, T. T. and P. Hunkeler. 1971. Evidence of tool-using by chimpanzees in the Ivory Coast. *Folia primatol.* 15:212–219.

Struhsaker, T. T. and L. Leland. 1979. Socioecology of five sympatric monkey species in the Kibale Forest, Uganda. *Advances in the Study of Behavior* 9:159–228.

Strum, S. C. 1975. Primate predation: Interim report on the development of a tradition in a troop of olive baboons. *Science* 187:755–757.

Sugiyama, Y. 1968. Social organization of chimpanzees in the Budongo Forest, Uganda. *Primates* 9:109–148.

Sugiyama, Y. 1969. Social behavior of chimpanzees in the Budongo Forest, Uganda. *Primates* 10:197–225.

Sugiyama, Y. 1972. Social characteristics and socialization of wild chimpanzees. In F. E. Porier, ed. *Primate Socialization*, pp. 145–163. New York: Random House.

Sugiyama, Y. and J. Koman. 1979. Social structure and dynamics of wild chimpanzees at Bossou, Guinea. *Primates* 20:323–339.

Sutter, J. 1958. Recherches sur les effets de la consanguinite chez l'homme. *Biologie Medical* 47:563–660. (Cited in Dobzhansky 1962.)

Suzuki, A. 1969. An ecological study of chimpanzees in a savanna woodland. *Primates* 10:103–148.

Suzuki, A. 1971. Carnivority and cannibalism observed among forest-living chimpanzees. *J. Anthrop. Soc. Nippon* 74:30–48.

Taylor, C. R., K. Schmidt-Nielson, and J. L. Robb. 1970. Scaling of energetic cost of running to body size in mammals. *Am. J. Physiol.* 219:1104–1107.

Teleki, G. 1972. The omnivorous chimpanzee. *Scientific American* 228(1):33–42.

Teleki, G. 1973. Group response to the accidental death of a chimpanzee in Gombe National Park, Tanzania. *Folia primatol.* 20:81–94.

Thomas, W. M. 1959. *The Harmless People*. New York: Knopf.

Tinbergen, N. 1956. On the functions of territories in gulls. *Ibis* 98:401–411.

Trivers, R. L. 1971. The evolution of reciprocal altruism. *Quart. Rev. Biol.* 46:35–57.

Trivers, R. L. 1972. Parental investment and sexual selection. In B. Campbell, ed. *Sexual Selection and the Descent of Man, 1871–1971*, pp. 136–179. Chicago: Aldine Atherton.

Trivers, R. L. 1974. Parent-offspring conflict. *Amer. Zool.* 14:249–264.

Turnbull, C. 1961. *The Forest People: A Study of the Pygmies of the Congo*. New York: Simon and Schuster.

Tutin, C. E. G. 1975. Sexual behavior and mating patterns in a community of wild chimpanzees (*Pan troglodytes*). Ph.D. dissertation, Edinburgh. (Cited by Short 1979.)

van Hooff, J. A. R. A. M. 1973. A structural analysis of the social behaviour of a semicaptive group of chimpanzees. In M. von Cronach and I. Vine, eds. *Social Communication and Movement*. London: Academic Press.

Van Orsdol, K. G. 1979. Slaughter of the innocents. *Animal Kingdom* 82(6):19–26.

Vayda, A. P. 1976. *War in Ecological Perspective: Persistence, Change and Adaptive Process in Three Oceanian Societies*. New York: Plenum.

vom Sall, F. S. and L. S. Howard. 1982. The regulation of infanticide and parental behavior: Implications for reproductive success in male mice. *Science* 215:1270–1271.

Waser, P. M. 1975. Monthly variations in feeding and activity patterns of the mangabey *Cercocebus albigena* (lyddeker). *E. Afr. Wildl. J.* 13:249–263.

Waser, P. M. 1980. Polyspecific associations of *Cercocebus albigena:* Geographic variation and ecological correlates. *Folia primatol.* 33:57–76.

Waser, P. M. and O. Floody. 1974. Ranging patterns of the mangabey, *Cercocebus albigena*, in the Kibale forest, Uganda. *Z. Tierpsychol.* 35:85–101.

Washburn, S. L. and I. DeVore. 1961. Social Behavior of baboons and early man. In S. L. Washburn, ed. *Social Life of Early Man*, pp. 91–105. Viking Fund Publications in Anthropology, no. 31, Wenner-Gren Foundation.

Watson, R. M., A. D. Graham, and I. S. C. Parker. 1969. A census of the large mammals of Loliondo controlled area, northern Tanzania. *E. Afr. Wildl. J.* 7:43–59.

Wiley, R. H. and D. G. Richards. 1978. Physical constraints on acoustic communication in the atmosphere: Implications for the evolution of animal vocalizations. *Behavioral Ecology and Sociobiology* 3(1):69–94.

Williams, G. C. 1966. *Adaptation and Natural Selection: A Critique of Some Current Evolutionary Thought.* Princeton, N. J.: Princeton University Press.

Williams, J. G. 1964. *A Field Guide to the Birds of East and Central Africa.* Boston: Houghton Mifflin.

Wilson, E. O. 1971. Competitive and aggressive behavior. In J. F. Eisenberg and W. Dillon, eds. *Man and Beast: Comparative Social Behavior*, pp. 183–217. Washington, D. C.: Smithsonian Inst. Press.

Wilson, E. O. 1973. Group selection and its significance for ecology. *Bio Science* 23:631–638.

Wilson, E. O. 1975. *Sociobiology, the New Synthesis.* Cambridge: Harvard University Press.

Wilson, E. O. 1978. *On Human Nature.* Cambridge: Harvard University Press.

Wing, L. D. and I. O. Buss. 1970. Elephants and forests. *Wildl. Monogr.* No. 19.

Wolfe, J. B. 1936. Effectiveness of token rewards for chimpanzees. *Comp. Psychol. Monogr.* 12:1–72.

Woodburn, J. 1968. An introduction to Hadza ecology. In R. B. Lee and I. DeVore, eds. *Man the hunter*, pp. 49–55. Chicago: Aldine.

Wrangham, R. W. 1974. Artificial feeding of chimpanzees and baboons in their natural habitat. *Anim. Behav.* 22:83–93.

Wrangham, R. W. 1975. The behavioural ecology of chimpanzees in Gombe National Park, Tanzania. Ph.D. dissertation. Univ. of Cambridge.

Wrangham, R. W. 1979a. Sex differences in chimpanzee dispersion. In Hamburg and McCown 1979:481–490.

Wrangham, R. W. 1979b. On the evolution of ape social systems. *Social Science Information* 18:335–368.

Würsig, B. 1979. Dolphins. *Scientific American* 240(3):136–148.

Wynne-Edwards, V. C. 1962. *Animal Dispersion in Relation to Social Behavior.* Edinburgh: Oliver and Boyd.

Wynne-Edwards, V. C. 1971. Intergroup selection in the evolution of social systems. In G. C. Williams, ed. *Group Selection*, pp. 93–104. Chicago: Aldine-Atherton.

Yengoyan, A. A. 1968. Demographic and ecological influences on aboriginal Australian marriage sections. In R. B. Lee and I. DeVore, eds. *Man the hunter*, pp. 185–199. Chicago: Aldine.

Yoshiba, H. 1964. Report of preliminary survey of the orang-utan in North Borneo. *Primates* 5:11–26.

Zahavi, A. 1974. Communal nesting by the Arabian babbler. *Ibis* 116:84–87.

INDEX

(All entries pertain to chimpanzees unless otherwise noted. Other species of mammals and birds are listed under their scientific names.)